D0433863

The Gardener's Book of Pests and Diseases

Roland Fox

The Gardener's Book of Pests and Diseases

BT Batsford

To Alicia,
who concentrates her efforts on growing her plants,
despite all the pests and diseases in our garden,
and a love of cats.

A CIP catalogue record for this book is available from the British Library.

ISBN 0 1734 7492 0

Printed in Singapore

For the Publishers

B.T. Batsford
583 Fulham Road
London SW6 5BY

Designed by DW Design Ltd, London

Photographs courtesy of Holt Studios International
and East Malling Research Station (pg 63)

Previous page: Grey mould spotting on rose.

Contents

Introduction

Ensuring a healthy start by creating a healthy garden

Gardens can be designed to exclude disease and pests from the onset. If this task is shirked the penalty is severe. Perpetual watchfulness will then be needed to counter the plagues of hungry pests and a variety of epidemics of withering disease.

Gardens are highly complex artificial environments that have been designed by humans. While all plants eventually fade, wither and decay in time, every keen gardener attempts to delay the process of decay as long as possible. In gardening, as in other spheres, success often relies on a combination of luck and good judgement to ward off the ravages of time and nature. Apparent miracles are possible. We all know of gardens that appear in good health and others where nothing grows well except the bill for new plants which never thrive. Just like an athlete, it pays to start good practices right from the beginning. Apart from disease organisms and pests that blow in from time to time, the health of a garden is determined when the soil is prepared for the first seeds that are sown, and the first trees or shrubs that are planted. The consequences last until remedial action has to be taken; in the meantime poor crops will be harvested and the garden will not flourish.

Sensible planning is essential in order to grow plants to their maximum potential, which will remain beautiful and healthy. Most pests and diseases reduce quality, some may occasionally cause the complete destruction of certain plants if detected too late. In these circumstances it is essential to detect outbreaks at the earliest stage and start preparing control measures immediately infection by disease or infestation by pests has been seen in neighbouring gardens or if damage is anticipated.

The extent of the injury that is likely to result to a plant from infestation by an unidentified pest or disease cannot be reliably forecast until it has been accurately diagnosed.

Although keys can be used, most gardeners may prefer to study the illustrations of the pests and diseases of major garden plants. Examples of the most damaging pests and diseases of key garden plants are illustrated in the encyclopaedic section of this book.

If a particularly destructive or unusual pest or disease is suspected, the causal agent must be quickly and accurately identified, so that appropriate action can be taken. Some diseased plants may even have to be eradicated under a Plant Health Order; many of these are indicated in the illustrations. More often an appropriate control measure has to be selected quickly if damage is to be minimised.

***Left*: Bud Blast on bud.**

Recognising plant diseases

The principles of plant pathology

Anticipating disease

Nearly all outbreaks of disease result from one or more of four major sources of infection, regardless of whether the pathogen is a fungus, virus or bacterium: (1) seed, (2) other planting material (seedlings, bulbs, corms, cuttings grafts etc), (3) airborne disease pathogens, (4) soil-borne disease pathogens. Seed that is available commercially must be tested to ascertain that it is free of pathogen contamination. Although heavily infected seeds can reveal direct evidence of pathogens such as bits of fungi or gummy masses of bacteria, most seeds rarely show any clear symptom of disease. In the average garden, the extent of many soil-borne diseases only becomes apparent over a number of seasons as they are rather slow-growing. Often this historic perspective is lacking and a range of diagnostic techniques (covered in Fox, 1993) is necessary to detect their presence in the soil before planting.

Once the amount of soil-borne disease is known, airborne pathogens form the main threat. If the arrival of showers of spores onto plants is correctly anticipated, protectant sprays and other control measures can be synchronised with them for best control. Several fungi as well as bacteria hitch a ride on a wide range of insects that feed on plants or their nectar. Some other fungal as well as bacterial pathogens, and many viruses arrive within airborne insects and mite vectors that can be controlled by insecticides or acaricide sprays.

Detecting early stages of disease before it can cause too much damage

Ever since the middle of the 19th century, scientists have paid considerable attention to discovering the entire details of the life cycle of the organisms that cause disease. As a consequence, it has been possible for plant pathologists to predict severe levels of diseases from past outbreaks. Latent infection is now recognised to be common among many diseases which become apparent after harvest, such as soft rot (*Erwinia carotovora*) and gangrene (*poma exigua var. foveata*) on potatoes, neck rot of onion (*Botrytis allii*) and grey mould of strawberries (*Botrytis cinerea*). As a result of this knowledge, recommendations for treatment for many of these diseases have been improved by enabling the application to be carefully targeted at a particular growth stage of the plant but only if a gardener is observant and well trained enough to recognise the presence of the pathogen.

***Left*: Close up of unidentified 'target' like fungal leaf spot on begonia leaf.**

What is disease?

Frequently a plant may not grow as well or as quickly as we would hope, or it may even die, because it is not functioning properly. This malfunctioning can be caused by a great variety of causes. Many insects or their larvae bite and burrow into leaves, stems and fruits. Water or minerals may be lacking or present in excess. The plant could be a casualty of pesticides or pollution or even be struck by lightning!

Strictly speaking, a plant only has a 'disease' whenever its malfunction is caused by a virus, a fungus or a bacterium. Other malfunctioning should be considered as a disorder. An expert in plant disease, known as a plant pathologist, combines the skills of your general practitioner as well as a public health official. In essence, a plant pathologist is a plant doctor. The diseases in the main section of this book have been categorised as per the principal part of the plant that is affected, such as leaves, fruits, roots, etc. However, you can sometimes find similar symptoms on several different parts of a plant. In each of these sections the English name of the disease is listed alphabetically and there is a very comprehensive cross index based on the plant, its Latin name, common name of the disease and the Latin name of the pathogen. Sometimes the same or a similar type of causal organism can have similar effects on a number of different host plants and may also require similar treatment.

The biology of plant pathogens

In order to understand why your plants become diseased we will briefly consider some features of the general biology of fungi, bacteria and viruses as pathogens. First, none of these organisms are themselves plants. Virtually all real plants contain chlorophyll which enables them to synthesise the organic nutrients required for energy and growth from sunlight, carbon dioxide in the air and water absorbed from the soil. Although some older textbooks link fungi to the plant kingdom, most fungi are now considered to be more closely related to animals and do not usually cause disease. Instead, they digest dead organic matter through which they grow by secreting enzymes and then absorbing the resulting substances. Man has learned to consume many of these fermented products – like the oriental tofu, blue cheeses, soy sauce, beers, wines and cider – but to shun those fruits and vegetables contaminated by moulds and rots.

Many fungi are edible and are grown commercially on compost or collected from the wild. Some of the latter, like ceps, chanterelles and truffles, grow in association with green plants as *mycorrhiza* but many woodland fungi decay leaf litter and fallen wood as saprophytes (organisms living on dead matter of all sorts). Some specialised fungi live on airplane fuel, and similar unusual sources of carbon compounds and other essential elements.Although they include hundreds of species that are important in causing garden diseases, a relatively small percent of all fungi attack plants. Certain pathogens are able to live on either dead or living matter (facultative pathogens). Some of these facultative pathogens can live on dead and dying organic matter, then use this as a base from which to attack healthy plant material. Even fewer fungi are actually completely (or very nearly totally) dependent on living plants as parasites; these are known as obligate pathogens. However, many obligate pathogens, such as the rusts and powdery mildews, are extremely common and damaging.In fact the organisms generally thought of as fungi are actually members of several distinct kingdoms of their own! Of these, only one resembles the plant kingdom by having a cell wall comprised of cellulose. However, like many animals, including man, these fungi – known as the Oomycetes – reproduce sexually through the fusion of motile cells that in some ways resemble sperm.

There are many hundreds of thousands of fungal species. Most, but not all, fungi consist not of single cells like so many plants and animals, but of microscopic tubular threads termed hyphae which are collectively known as a mycelium. Hyphae germinate from a variety of spores formed by sexual or asexual reproduction. Many varied fungi forms produce several different spores of distinctive types. These are formed by various characteristic processes.

These spores develop into either the well-known toadstools or numerous other smaller but equally intricate fungal constructions. Mycelium of many plant pathogens spreads out through the soil, decaying vegetation and plants. You may see it as a white cottony material. Many fungicidal chemicals work by controlling spore germination, but others can control the hyphae within plants. The former are called protectants, whereas the latter are systemic eradicants (these eradicants are said to possess 'kickback' action).

Generally the asexual state of the fungus causes the disease. This is given a different name to the usually less common sexual stage. Recognition of the particular type of spore that is produced is vital in traditional fungal taxonomy. Some fungi, the *Deuteromycotina*, sometimes called the *fungi imperfecti*, produce only asexual spores called conidia. However, if such fungi are later found to undergo sexual reproduction to make ascospores in sac-like structures; they are then transferred to the Ascomycotina. The majority of fungi belong to this class. Fungi in the class *Basidiomycotina* – like mushrooms or toadstools, rusts and smuts – produce exposed sexual spores termed basidiospores. Other spores are produced in the rusts and other fungi.

Pseudofungi include the Oomycetes, such as the causal agents of potato blight, downy mildews and damping-off, and similar organisms such as those responsible for potato wart have free-swimming zoospores which move through films of water both in the soil and across the surface of plants. Clubroot, although caused by a taxonomically distinct organism, also produces zoospores.

Fungi persist under adverse conditions as mycelium or as especially resistant spores or other impenetrable structures such as tough masses of mycelium (*sclerotia*). Once these are controlled the life cycle of the pathogen is destroyed.

Bacteria are different

Very few plant pathogenic bacteria produce spores. Actinomycetes like *Streptomyces scabies* that causes common scab form powdery spores from the tips of their chains of vegetative cells. Most bacteria exist as individual minute single rods or round cells that multiply by simple binary fission, as the cells split into two with remarkable speed, especially under tropical temperatures. Many bacteria are capable of movement through fluids. The conspicuous symptoms and other effects of bacterial infections often resemble those produced by fungi. The presence of bacterial pathogens is often difficult to corroborate as bacteria can be found virtually anywhere as saprophytes.

Bacterial diseases often prove difficult to control once they are established. Fungicides are seldom effective against bacteria.

A number of extremely important bacterial diseases regularly occur in gardens in Britain. Fireblight caused by *Erwinia amylovora* is one of the most serious diseases that can be found on cotoneaster, pyracantha, *Sorbus*, apples, pears and many other pomaceous plants. Soft rot of vegetables and some fruits is caused by another related bacterium, *Erwinia carotovora*. The damage that this causes is likely to be all too familiar to your sense of smell! Other bacteria, often from the tropics, are very important as pathogens on many glasshouse plants.

Recognising symptoms

It is essential that a plant disease has been correctly identified before any control is attempted, as some treatments designed to destroy one pathogen may encourage others. However, while the signs or symptoms of some common pathogens are familiar and may be trusted, those of many others frequently prove less reliable as a guide to the identity of the pathogen. Nevertheless, when clearly distinct, symptoms can often help to monitor the spread and prevalence of diseases on a range of garden plants.

***Left*: Mealybug colony on houseplant.**

Symptoms as evidence of disease

Most symptoms are the visible outward signs of a distinct malfunctioning in a host plant brought about by the action of harmful pathogens or other causal agents such as mineral deficiency or pollution. Symptoms can range from extremely slight to very serious, chronic to acute. They may be localised, resulting in restricted injuries like leaf spots. Some are systemic, spreading throughout an affected plant to cause widespread symptoms such as distortion or rot.

Obligate fungal pathogens depend on haustoria, which are specialised hyphae that drain nutrients from host cells without killing the plant. This is a successful strategy as the host survives virtually intact. If the attack were to be too extensive the host might be killed, followed either by the death by starvation of the obligate pathogen or the hasty formation of resistant spores.

Facultative pathogens are usually considerably more injurious. At first they grow amongst host cells, but often later they penetrate within, so as the lesion advances it causes substantial damage at the cellular level. This results from a series of reactions between two components, the basic host cell functions (photosynthesis, respiration and transport) and activities of the pathogen (toxins, enzymes and growth regulators). Although a conspicuous symptom is a clear confirmation of a detrimental interaction, many symptoms are not readily discernible since they occur at the cellular level.

However, as soon as these micro-symptoms occur *en masse.* a macro-symptom may result. This is clearly visible to the naked eye. Although the onset of perceptible changes in host function – or morphology – may ultimately become evident by the development of a particular micro- or macro-symptom of the condition of the disease, sometimes the most obvious symptom of a plant disease consists of the vegetative or fruiting structures of the pathogen itself, often termed 'signs'. Other manifestations of disease require a sense of smell, taste or touch.

Use of macro-symptoms in disease detection

Inspecting plants for macro-symptoms may be relatively rapid. However, the appearance of any symptom represents a relatively late stage in the process of infection and colonisation by a pathogen. Yet the unexpected appearance of a symptom is usually sufficiently conspicuous to worry a keen gardener into identifying the causal agent.

Use of symptoms for diagnosis

Experienced gardeners can sometimes accurately diagnose a common disease from its symptoms even when seen from a fast car. However, in most cases a rather more detailed examination is required as plants are covered by many micro-organisms that are unable to attack plants.

Drawbacks of depending on the use of symptoms for disease diagnosis

A symptom can be so distinctive that it confers the common name of the disease. Often several pathogens cause similar symptoms. Symptoms range in severity; some are extremely destructive, others hardly seem worth treating, but even slight damage become important if persistent. The loss of a single leaf is trivial, but when all the leaves are lost the plant can die. If a pathogen rots the roots of a plant, permanent wilting may follow. Similar symptoms result from any blockage to the vascular system, for example, when filled by a wilt disease. Alternatively its leaves could have been severely damaged by a foliar pathogen.

Symptoms caused by agents other than pathogens

From the outset it is essential to make sure that the symptom is due to a pathogen and not caused by an insect, nematode, a variety of other pests, parasitic plants, some weeds, microbial toxins or a range of soil or climatic conditions. Symptoms that look very much like many diseases may be produced by excessive imbalances in minerals, exceptionally high or low temperatures, light, water supplies, chemicals, lightning and wind. Non-infectious disorders are therefore often mistaken for diseases induced by pathogenic fungi, bacteria and viruses. Both affect the plant in similar ways. Now and then disorders allow pathogens to invade and injure the plant. This hinders diagnosis. Damage differs with different plants, their maturity, the season and the different parts of the plant. Many garden plants can survive very low temperatures and frost, yet freezing may injure or kill twigs and branches and even split trunks after heavy frost. Fruit crops may be lost when flowers are killed in late frosts and some plants, such as dahlia, can even be killed by early frosts before flowering. In cases of sunscald, fruit or foliage is injured by bright sunshine. Heat cankers of the collar and black heart of potato are caused by temperatures at which the rate of respiration of oxygen is so high that it is used up too fast to be replenished. Poor light causes etiolation, but an excess may induce sunscald of bean pods. Drought

is often a problem. Maize starts to roll its leaves if short of water; if the drought continues its top dries up so that grain cannot form. Trees may die in prolonged droughts. Most garden plants thrive on relatively well-drained soil, but cannot survive persistent floods that suffocate the root system, allowing micro-organisms to invade.

A number of apple varieties suffer badly from scald, a serious physiological disorder incited by the gases that they emit in storage. Yellowing and other colour changes in various patterns in the leaves, marginal scorching and other forms of necrosis, as well as stunting or deformation of the fruit may indicate that one or more of the elements in the soil essential for plant growth is deficient or has combined with other soil elements making them unavailable to the plant. More rarely there is an overabundance of salts in excessively acid or alkaline soils.

Symptoms of disease are conspicuous when some of the major fertilizer elements such as potassium, phosphorus or nitrogen are lacking. Damage is most often marked when some of the minor elements such as boron, zinc and manganese are missing in the soil or are rendered unavailable. An overabundance of some nutrient elements, such as boron, in irrigation water or as a contaminant in potash fertilizers, results in a marginal necrosis of the older leaves. Occasionally mineral excess causes stunting, resulting in serious losses and even dead plants. It is often difficult in practice to distinguish clearly between the symptoms of disorders that are due to a deficiency and those due to an excess. Apparent shortages of one element are often caused by a surplus of another element that interferes with the solubility, absorption or function of others.

Pesticide sprays may sometimes cause phytotoxic blemishes on the fruit or foliage. The most common injury is a dull brown spotting of the leaves or burning of margins and tips. Even minute amounts of the growth regulator herbicides can deform or kill certain plants like tomatoes. Other consequences of drift of spray droplets include poor growth, excessive drought injury and premature flower or fruit drop.

Classification of major types of disease symptoms

Pathogens can cause a variety of immediate or delayed injuries to plant functions following infection. These plant functions (together with the groups of diseases affecting them) include the mobilisation of stored food (damping-off and seedling blights), absorption of water and minerals (root and foot rots), water transport (vascular wilts), meristematic activity (leaf curl, witches' broom, club-root, galls), photosynthesis (leaf spot, anthracnose, blight, mildew, rusts, leaf smuts, viruses), translocation (some

***Left*: Grey mould disease on paeony flower buds.**

diseases caused by viruses, viroids, mycoplasmas, rickettsias), storage (post-harvest diseases of fruits, perennating and storage organs) and reproduction (head smuts and ergot). Various symptom types may be grouped together on the basis of these distinct underlying mechanisms. Diseases usually have several symptoms which together constitute a syndrome, some of which are so distinctive that a disease can be readily identified and controlled.

Controlling plant diseases

The effective control of plant diseases can be ensured in several different ways. Many of these methods do not require the use of chemicals even though this is often the usual procedure adopted by amateurs. Only a few of the fungicides that are available to professional growers can be used in the garden. Some of these products are systemic and can be applied either to the roots or leaves to eradicate diseases that have already become established in younger parts of the plant. Unfortunately, although such fungicides are extremely effective initially, prolonged use in the absence of other treatments can select strains of some pathogens that are resistant to the systemic fungicide. The other fungicides that are available are protectant and must be applied to plants before infection takes place in order to establish a protectant barrier between the host and the causal agent of disease. It is much less common for resistant strains to develop into most standard protectant fungicides. Most fungicides available amateurs to use only control foliar diseases adequately. Nonetheless, it is also possible for an amateur to purchase seeds which have already been treated with small amounts of a systemic or protectant fungicide by the seed company. This fungicide is intended to control any seed-borne pathogens and in addition give some protection against soil-borne diseases such as damping-off. Most garden fungicides are wettable powders which consist of finely ground or precipitated particles that remain in suspension in the water used to spray them but do not dissolve. In addition to between 25 and 50 percent of the active ingredient, wettable powders also contain a wetting agent, a thickening agent, a clay and a suspension agent. Most fungicides applied in the garden are sprayed through the nozzle of a simple hydraulic aerosol sprayer of some kind, whereas commercial growers use a much wider variety of equipment. Many gardeners are not keen to use fungicides for various reasons including the fear that they might harm the environment, the dislike of handling pesticides themselves, the cost of purchasing spray equipment and the desire to use more traditional techniques. Fortunately there are a number of effective methods of disease control that do not rely on the use of chemicals at all, as well as others that integrate their use into systems that reduce reliance on them. Integrated disease management is at present largely used by professional growers, but elements of this philosophy can be adopted in the garden. These systems are based on the idea of containing damage or loss below certain economic levels. This aim is achieved by the management of the growth of plants by several processes in which disease is only one component. As a result there is less dependence on chemicals and hence a reduction of possible detrimental effects on non-target, and possibly beneficial, organisms. There are more of these techniques available than is generally realised by amateur gardeners.

One of the most important lessons that any gardener can learn is not to introduce diseases into the garden. It is very easy to purchase diseased plants from charity plant sales or receive them from well-meaning friends and neighbours. Among the diseases spread in this way are a number of serious, soil-borne pathogens including honey fungus root rot of trees, which has been disseminated in this way on a number of occasions to new gardens and has then been very difficult to eliminate. On a larger scale so many diseases have been transmitted between countries in the same way that international legislation by regional plant protection organisations of the Food and Agriculture Organisation has been set up to prevent the accidental spread of diseases such as downy and powdery mildew of grapes and powdery mildew of gooseberries – which were introduced into Europe from North America – and Dutch elm disease, apple canker and black leg of crucifers which spread in the opposite direction. In addition to the international spread of diseases, some of which were possibly imported by keen gardeners who disobeyed quarantine laws in Britain, some serious diseases such as potato wart are localised and there is legislation to prevent further spread. There has also been similar legislation, albeit ultimately unsuccessful, against the spread of Dutch elm disease, fireblight and white rust of chrysanthemums. Also, it is illegal to offer for sale plants affected by gooseberry powdery mildew, onion smut, cabbage club root, red core of strawberry, progressive wilt of hops or even the pear variety 'Laxton Superb' which is known to be especially susceptible to fireblight. The important message here is to make sure that you only add healthy plants to your garden and avoid the temptation to bring in any plant that is not in the best condition. This would therefore rule out the introduction of many wild plants. It is most important to obey plant health rules, however strict they appear to be, as once a disease is introduced it is very costly to eradicate it – if this can be done at all. Only a few cases of the successful eradication of a plant disease have been reported; the most effective was the complete elimination of citrus canker from florida and the Gulf of Mexico, and that was at the cost of 20 million trees.

Eradication of diseases from the soil, seeds, vegetative parts or non-living surfaces is an effective way of cleaning up plants and their environment. In many cases chemicals are used, but eradication by physical means such as heat is also widely used. Heat can be applied in the form of thermal radiation from a heating element, fire used to burn garden debris, radiation from the sun, ionizing radiations or microwaves, live or pressurised steam, or hot water. The use of heat is generally limited to small areas such as greenhouses, nurseries and small transplant seedbeds, but is rarely used by amateur gardeners. Professional seed and bulb producers have used heat, particularly hot water, to disinfect seeds, bulbs and corms, but the gap between success and damage is often too small for this technique to be worthwhile. Even though it is quick to use and usually leaves no harmful residues, heat is not easy to handle and destroys organic matter in the soil, and burning causes air pollution. Another adverse factor is that soil which has been heat-sterilized is easily reinfected by soil-borne pathogens. It is very difficult to sterilize soil or compost completely. Often soil or compost has actually been pasteurized, that is, it has been heated to 62°C. This kills the more sensitive organisms such as most plant pathogenic nematodes, many bacteria and fungi such as *Pythium*. Moist heat is more efficient than dry heat in eliminating pathogens as water conducts heat more readily than air. Steam readily denatures the proteins and membranes in micro-organisms for fewer heat units than dry heat. In hot countries it is common for patches of soil to be disinfected by heat from the sun; now a development of this technique involves the use of a polythene sheet to heat the soil to as much as 55°C to a depth of 5cm. This technique is known as solarization, but it will probably remain restricted to tropical and Mediterranean areas. Kitchen microwave ovens have been regularly used to disinfect small batches of soil and composts, but this method is not always successful because the uneven texture of some soils prevents complete pasteurization.

Other methods of disease control make use of our ability to modify the environment and tip conditions to favour the host plants rather than the pathogen. This is really a form of plant health management by encouraging vigour rather than applying chemicals as protectants or eradicants. The mechanisms by which cultural practices reduce disease development, its severity or both include disease escape, indirect biological control, eradication or exclusion. Many pathogens such as bacteria, rust and mildew fungi attack young succulent tissue before it ages and hardens. Numerous other pathogens thrive on weak or senescent tissues on plants that lack vigour. Hence the way that the environment is modified depends on the type of pathogen that is known to be present. For example, increasing the levels of nitrogenous fertilizers used on apple trees will reduce apple scab but increase the damage caused by fireblight. Overhead watering can often reduce losses from crown and root diseases but increase leaf diseases. The presence or absence of moisture is probably the most important requisite needed by pathogens before a disease can develop. Three forms are involved with disease, free water on the surfaces of plants, high relative humidity and soil moisture. Dangerously high levels of these can be altered by cultural practices including reducing the amounts of water used in irrigation, reducing plant density and improving drainage. However, not all diseases thrive in wet soils and there are a number that favour drier conditions. Free moisture from rainfall or overhead watering promotes the development and spread of a number of leaf diseases. While it is possible to shelter the leaves of peaches and almonds from rain with a polythene cover and thus reduce leaf curl disease, this is a relatively unusual technique, even in Britain where such foliar diseases are very common. Nonetheless it is much easier to water plants from below and avoid exacerbating leaf diseases. Alternatively, if this is not feasible because of uneven surfaces it is possible to continue to water overhead but to do it less frequently. Another technique to reduce leaf wetness is to water early in the day rather than late afternoon. It also pays to avoid watering during cloudy weather as this also prevents prolonged periods of wet foliage. The use of misting in greenhouses and propagation chambers is also likely to make foliage diseases worse. As well as making conditions suitable for disease development, overhead watering and misting also allows fungal spores, bacterial cells and nematodes to be spread between plants by splashing from infected plants or soil onto healthy plants. Many diseases are most severe in poorly drained soil; improved plant growth will result if the soil is drained using ditches, tiles or pipes. Another way of ensuring that plants are well drained is to grow them on raised beds, ridges or mounds. In general flooding should be avoided, but there are some pathogens like certain wilt fungi that are killed either by too much water or are stressed and become attacked by antagonists that can survive under wet conditions. However, in general regularly flooded soil is often low in beneficial organisms.

Planting plants relatively wide apart reduces both relative humidity and also free moisture on the foliage; this in turn diminishes the risks from powdery mildew and grey mould.

After moisture, the most important factor that influences disease development is temperature. This can also be affected by a number of cultural practices. One of the most fundamental influences is the date of planting seeds or bulbs outdoors. Seedlings planted later in the spring grow much more rapidly and hence are susceptible to many soil-borne diseases for a far shorter time than those that have to contend with less favourable conditions.

The temperatures of both plants and the soil beneath them can be reduced by shading, but this can also increase moisture, especially if the relative humidity of the trapped air becomes too high. Shading from fences, buildings, trees or other plants, such as rank weeds, increases infection caused by powdery mildews. Removing the shade and allowing the temperature to rise reverses this effect, possible also by promoting the growth of the hosts. Some soil-borne diseases such as root rot of rhododendron can be reduced by maintaining soil temperatures above 33°C. Ventilation by pruning and thinning, hence allowing air to circulate freely around plants which are well spaced out, decreases the air temperature and humidity and tends to reduce many diseases compared with conditions under dense plantings. Pruning and thinning are also important ways in which diseased plants and parts of plants are removed from otherwise healthy populations. The orientation of rows of plants also helps to dry and cool plants. Aeration of the soil has a more direct effect on the survival and behaviour of soil micro-organisms including pathogens. Most plant pathogens in the soil require some air to live and tend to survive best in the upper layers of the soil. Digging the soil breaks it up and creates aeration through the air channels that are created. The deeper the digging that takes place the better the subsequent root growth, as the roots can often penetrate into soil layers where there are fewer pathogens. Drainage is also improved and corrects the effects of waterlogging that encourage some fungi and nematodes.

The relative acidity or alkalinity of the soil influences the types of plant diseases that occur and so it is possible to discourage them by adjusting the pH of the soil by adding lime to raise it or some inorganic fertilizers, such as those based on ammonia, to lower it. In the same way a number of organic amendments, such as animal manures or composts based on plant material can provide nutrients, improve soil structure and have a beneficial effect on the population of soil micro-organisms. Among the plant materials that are incorporated into soils are green plants of various kinds, straw, wood chips and tree bark. As these do not contain much nitrogen this should be applied in addition. The soil around acid-loving plants is often amended with sphagnum moss peat as this will reduce the pH as well as improve water holding capacity.

Another technique which can isolate plant pathogens is to dig a trench between the infected plants and their healthy neighbours. Some success with this method has been reported in preventing the root-to-root spread of Dutch elm disease and several other tree diseases that spread through the soil. However, the use of trenches or barriers of brick or metal has been less successful in preventing the spread of honey fungus as its rhizomorphs can follow the roots down deeply into the soil well below the level at which these barriers are feasible. Other barriers – screenhouses – have been set up to prevent aphids spreading viruses to plants, but these are too expensive and unsightly for the ornamental garden.

It has been found that there is usually a limit to which pathogens can spread from diseased plants, but as this can vary from a few metres to several kilometres this information is rarely of much use, although it is generally sensible to site plants upwind from known sources of infection – for example, from neighbouring gardens. Windbreaks, both artificial and natural, protect plants from damage and also affect the dispersal and deposition of wind-blown inoculum as well as aphid-borne viruses. Many different types of diseases can be reduced in this way, including fire blight and other bacterial diseases. Other diseases such as rusts, some leaf spots and virus diseases are often more prevalent to the leeward of windbreaks as quiet air in this area allows pathogens and their vectors to settle out (rather as snow drifts settle in the same places), but beyond this zone lies an area that is relatively free of infection. This disease-free zone is, however, not very deep and unless there are a number of windbreaks their effect is not significant. Nevertheless, in many gardens there are often a number of trees, bushes and other natural features that can act as windbreaks so the cumulative effects may be important.

Mulches – whose prime function is to conserve or increase soil moisture or temperature and control weeds by forming a physical barrier on the soil surface – can have a beneficial effect by preventing pathogens present in the soil from splashing up to the foliage.

Some bacterial diseases can be aggravated if the hosts are planted in a frost hollow. It is possible for extra damage to be caused by the bacteria functioning as centres of ice formation and ice nucleation, and it is actually this process which disrupts the plant tissues. Therefore when planting cherries and other susceptible plants make sure that cold air drains off the site freely.

Another way of reducing the level of disease is to remove or retard those tissues or organs through which the pathogens enter. For example, fireblight bacteria enter the flowers of their hosts when the temperatures are between 16 and 27°C, but if the trees or shrubs blossom earlier than this they will escape disease as the temperature will be too low for proper infection. Although some pears and other hosts are not normally vulnerable for this reason they can still be especially susceptible if they produce a secondary period of blossom that occurs when the temperatures are higher. Any secondary or late blossom should therefore be removed in order to reduce the risk of infection. In Britain it is now prohibited to grow these varieties of pear. Several diseases are affected by the timing of pruning that coincides with wet periods; if pruning is delayed until the weather is drier then some bacterial and fungal cankers are reduced.

Recognising damage by plant pests

Types of plant pests

Plant pests fall into the categories of those which occur every year and cause serious damage, those which are always present but cause little damage unless they occur in plague proportions during favourable weather, and those that may cause so little damage that control is unnecessary. It is necessary to understand the life history of pests in order to know into which category they belong and hence whether control is worthwhile and what measures are best applied.

There are several different types of damage caused by plant pests, depending on how they feed. The majority of insects, woodlice and millipedes feed on plants by biting and chewing the pieces of tissue with jaws that are specially designed for this purpose. Other insects, particularly the larvae of flies, have jaws that operate like hooks which tear off pieces to be digested in a mushy mass. Slugs and snails rasp away at plants with a tongue-like radula, covered with myriads of minute hooked teeth. This can be so effective that the mollusc can tunnel and burrow into fairly solid host tissue such as potato tubers or corms, as well as devastating leaves and flowers, often so completely that nothing remains.

Many pests eat the leaves of plants – among these are caterpillars, sawfly larvae, beetles including weevils, and slugs and snails. Other caterpillars and sawfly larvae strip the surface of one side of a leaf, leaving the other side to die along with the leaf veins. The loss of leaves reduces the efficiency of the plants, preventing growth, and in extreme circumstances can kill the plant. Other biting pests like caterpillars and beetles such as chafers eat buds, but this sort of damage can also be caused by birds, particularly bullfinches. The effect of this is to slow down growth of the main shoots, allowing dormant buds to burst, thus crowding out the normal development of the plant. This sort of damage can seriously affect some trees and greatly reduces flower production.

Several pests including caterpillars, earwigs and some beetles attack the flowers and reduce their ability to produce fruit and seeds properly.
A number of other insect pests such as caterpillars, leatherjackets, destroy underground stems and affect the conductive tissue in the stems of plants. This can cause serious injury as water and sap cannot flow properly resulting in reduced growth, wilting and even death. This can be especially damaging in trees that are affected by beetles which burrow into the area under the bark and destroy the cambial tissues.

The damage to plants caused by biting pests also allows disease organisms to enter plants. This effect is even more pronounced if the biting

***Left*: Rhododendron leaf hopper.**

pest has fed on diseased plant tissue and carries fungal spores or bacteria with it to new plants. Occasionally viruses are also transmitted by biting pests, but most are spread by sucking pests.

Sucking pests feed by piercing the tissues of plants to suck out the sap which contains proteins as well as abundant sugars. Often the sugars are jettisoned, excreted as honeydew – spots of which cover the leaves of the plants below the insect. Sucking pests include aphids, whiteflies, leafhoppers and capsid bugs which belong to the Hemiptera. In these, the mouth parts including the mandibles have become elongated into a needle-like stylet enclosed in a protective sheath. This enables these insects to pierce through the plant, inject digestive enzymes and suck out sap.

Thrips also wound the surface of leaves in order to suck up the sap, but have not evolved such effective mouth parts as the Hemiptera. Other pests that pierce plants to suck out sap include mites and nematodes. Nematodes, or eelworms, extract sap from plants by using a hollow mouth-spear to pierce the cells. But whatever means is used to remove sap from a plant, its vigour will be affected, growth is reduced, wilting may take place and the plant may die. flowers will be malformed and discoloured, or may not even form at all (a situation known as 'blindness') if the flower buds are attacked by capsids, other sucking insects or leaf nematodes. Other sucking pests such as red spider mites, thrips, leaf hoppers and whiteflies cause pale yellow mottling to the leaves on which they feed. The saliva of many sucking pests is toxic to plants, resulting in effects such as discoloured leaves caused by aphid feeding, curling of leaves and distortion of stems caused by aphids, capsid bugs or nematodes and unusual growths or swellings caused by adelgids, aphids, gall mites or nematodes.

As well as fungi and bacteria that can either enter with or after wounding by sucking pests, there are many viruses and mycoplasma that are spread by a variety of pests including aphids, leaf hoppers, thrips and eelworms. In order to recognise the presence of pests it is necessary to recognise the symptoms of damage caused by the pest as well as the pest itself, as often only the former may be obvious, since many pests are secretive or are only active at night. The encyclopaedic section of this book deals with both aspects, with a description of the damage seen on different parts of plants, as well as describing the appearance of the pest.

Controlling plant pests

In the garden it is very easy to consider chemical control alone to be the most feasible method of controlling pests. While it is true that there is a range of different insecticides available to the amateur – and many more for the professional – that will kill one or more of the important garden pests, they do have quite different properties and behave in a number of ways. The most obvious split is between compounds which are systemic and move within the plant, often from drenches applied to the roots, and those which are non-systemic. While non-systemic chemicals remain on the surface of the plant or stay in the soil to protect against pest attack, the systemic chemicals can eradicate pests that feed on plants or live within them. All insecticides should be handled with caution and the label put on the product by the manufacturer should give the recommended dose to be applied and the handling instructions, the product name and concentration, as well as the trade name. Often an instruction leaflet is also included – it is important to read both before attempting to use any product. Among the general rules that apply, one should wear disposible rubber gloves before handling the concentrate; always add the concentrate to the water and never add water to the concentrate; always wash hands and the concentrate bottle when you have finished and before spraying. For some products a face shield or safety glasses should be worn when pouring the concentrate, but even if this is not mentioned always look away slightly when you do this. Always make sure that you measure out and add the correct amount of concentrate to the amount of water recommended on the label by the manufacturer.

As well as the potential hazard pesticides pose to those who use them, some, particularly the non-systemic chemicals, also pose a threat to non-harmful and beneficial insects. It is therefore wise to spray insecticides only if it is really necessary. In fact, in many situations the mere presence of the pest may not in itself be a sufficient reason for spraying. If the numbers of the pest are insufficient to cause a substantial loss of either the plants themselves or their quality then it is often not worth the risk of killing the insects that can feed on the pests. For example, you may notice ladybirds on a plant among the aphids which, if left a couple of days, will eat small populations of the aphids without the need for spraying at all. You should never apply more insecticide than the dose that is recommended by the manufacturer on the label. In any case this will not ensure any better pest control and could seriously affect the survival of beneficial insects and the length of time harmful residues persist.

Another way of protecting beneficial insects is to use systemic insecticides. These are really only effective against sucking pests, particularly aphids. Aphids probe deeply into the plants to obtain sap and while they are doing this they pick up the insecticide. This will not affect harmless insects on the plant surface, especially if the chemical is applied to the soil around the plant to be treated rather than sprayed directly onto it. In this way not only do the leaves remain safe for other insects but the treatment often lasts longer than a spray. Granules can be used which are safer and more convenient to apply. The only drawback is that the soil drench or granule treatment will not be as immediate as a spray applied directly to the leaves. Another way of reducing the harmful side effects of a spray is to 'spot treat' only those plants or parts of a plant that are infested by the pest and not foliage that has yet to be colonised. In this way you can avoid contaminating the soil and plants where pests are absent but beneficial insects are present, often hidden away during the day if they are active at night.

Pesticides are usually formulated with adjuvants such as wetters and stickers. These are important additives which make it possible to apply the most appropriate amount of the active ingredient and prevent undue run-off and wastage. Occasionally additional wetters or stickers are necessary to treat plants with smooth, waxy or hairy leaves. The type of sprayer that is used also determines how efficiently sprays are applied. One of the most efficient sprayers does not use a pump to push spray through a nozzle but spins off a shower of appropriately even-sized droplets by centrifugal forces from a spinning disc within a cage. Simple 'controlled droplet applicators' are such devices, some of which operate from a two-stroke engine (similar to a strimmer). By contrast with most conventional sprayers and also aerosol cans which use pressure to force spray through a nozzle produce, more than half the spray is in the form of large droplets 1/3mm in diameter which will either bounce off the leaf or wet it so much that most of it runs off onto the soil. The objective of spraying should not be to wet the leaf obviously but instead to produce a film of much smaller droplets that are difficult to see with the naked eye unless the leaf surface is wiped. Even though little spray is wasted, there is a danger of losing chemicals if overly small droplets are produced as these do not have the weight to overcome air movement and thus will disappear up into the air to be deposited where they were not intended to go! Really only a narrow range of droplets about 0.1mm in diameter is needed as these size droplets will land and stay on the leaves. With a conventional sprayer typically only 1.5 percent of the total spray ends up as these useful droplets; the rest are wasted as they are either to fine or too large, representing an unnecessary waste of money and a source of contamination and hazard for the operator if there is too much wind. Some hand sprayers have a nozzle which can be adjusted to give fine droplets of an appropriate size. The best way of achieving optimum results with an adjustable-nozzle hand sprayer is by trial and error, making sure that the sprayed leaves still appear dry, but at the same time the presence of droplets can be demonstrated by smearing the leaf with a stick. If spray droplets can be seen they are too large and spray is being wasted.

Before chemical control was generally available many gardeners used a range of other techniques to control plant pests. Some of these techniques work well, but others have little or no effect when tested scientifically. For example, there is only a slight effect on yield if the tips of blackfly-infested broad beans are pinched out, but slipping a square of carpet around the stem of a cabbage definitely controls cabbage root fly. Other methods of controlling pests without resorting to pesticides have been developed over the years and are well worth using either instead of chemical control or to reduce the amount of spraying that is needed.

The simplest method of cultural control is to ensure that the plants are growing as vigorously as possible but without the excessively sappy growth associated with nitrogenous fertilizers which encourages aphids, leafhoppers and other sap sucking pests. It is also wise to eliminate weeds as this removes competition and allows better plant growth, removes alternative host plants for insects to feed on and it reduces the amount of shelter for pests like earwigs and capsid bugs. Cultivation itself is helpful as pests are either destroyed or brought to the surface where they are eaten by birds, freeze or dry out in the sun. Digging can also bury pupae so deep that they cannot emerge and reach the surface.

One technique, which is often used in many developing countries, is to mix plants rather than to plant large blocks of the same species. The greater the variety of plants that are grown the less chance that a pest has of finding an appropriate host. Insects arriving in an area with many different plants are confused by the variety of odours from plants on which they cannot feed. This is one of the reasons why pest epidemics are less common on wild plants than on their cultivated relations which are grown in large fields by farmers. It is also wise to maintain plenty of flowers to provide a supply of nectar to feed the natural enemies of pests; flowers also yield a source of protein for these insects to produce eggs. Among the predatory insects that feed on flowers are hover flies, assassin flies, predatory bugs, ladybirds and many parasitic insects.

Another way of avoiding plant pests is to avoid having plants at a susceptible stage when the pest is present. A good example of this is demonstrated by the pea moth which has to synchronise laying its eggs when its host peas are just finishing flowering so that the pods that the caterpillar needs are just ready for it to bore into when the egg hatches. Pea pods at other stages of growth are unsuitable for egg laying and so are not attacked. The secret of successfully growing peas without pesticides is therefore to ensure that the pea plants are either too advanced or too late for the pest.

When annual plants are grown it quite often the case that some sowings are much more successful than others, so it is well worth experimenting with different planting dates to find one that provides the best 'window of opportunity' for that particular plant. It is also possible to provide this window of opportunity by taking care with techniques that prevent the growth of plants to be checked such as deciding on suitable spacing, fertilisation and watering as well as choosing suitable planting and transplanting dates. In a similar way many soil pests, particularly root flies, are attracted to crops growing in bare soil. The reason for this is that many of these plants evolved in ecosystems based on bare soil and their pests need soft soil into which to lay their eggs. In the garden the act of thinning plants provides just the conditions that these pests need to give their larval maggot a suitable start in life in loose soil. The simple expedient of either not thinning at all, or changing the date of thinning by hastening or delaying thinning and thus providing windows of opportunity will reduce losses considerably.

Rotating the position of herbaceous plants is another well established routine that has a basis in science and can be shown to be effective. Although many pests such as aphids arrive in the garden each year as fresh infestations, many pests, particularly those that overwinter in the soil as a pupa emerge in the spring. If the position of their hosts is changed from year to year it is likely that the pests will have some difficulty in locating a suitable plant. Rotations for pest and disease control are widely used in agriculture for the same reason, eventually the 'reservoir' of soil-borne pests and diseases declines if suitable hosts are absent. In the wild many plants rely on the production of widely dispersed seedlings to avoid the build-up of pests and diseases that eventually destroy the parent plant and its offshoots which cannot escape from the original site. In a similar way many pests and diseases build up on fallen leaves around an infested tree. Many fallen leaves provide a snug shelter in winter for a number of pests, so it is wise to sweep these up and burn them as soon as possible in the autumn. The remains of stems and stalks in the herbaceous border provide similar lodging sites for a variety of pests during the harsh days of winter and so should be removed. Some aphids overwinter on plants that will sprout in the spring and these should be inspected thoroughly. Another well established method of controlling pests without pesticides or more frequently, using less pesticide input, is to choose a resistant variety. Different varieties of plants usually show different levels of resistance to varied pests often due to characteristics that are still unpredictable, so it can pay to read plant and seed catalogues to select pesticides if you have a particular pest problem, or if you cannot find this information to try several varieties of the plant you want since partial resistance can be discovered in this way. Sometimes this partial resistance means that a particular variety performs better in one garden than another, but this can only be discovered by trial and error, or observing which plants grow best in neighbours' gardens. Partial resistance is important as it can be very effective, resulting in an acceptable level of pest damage when combined with other pest control techniques such as biological control. Some varieties of garden plants have been specifically bred by specialist plant breeders with an especially high level of resistance against a particular pest; these varieties are usually mentioned in plant and seed catalogues and so are easily selected.

Biological control of pests is very important. In the wild and in many mature gardens there are many naturally occurring predators. You should learn to recognise these and not destroy them by mistake. In most gardens you would expect to find the following insects carrying out an effective job of biological control. A number of different ladybird larvae and adults are voracious devorers of aphids. Ground beetles come in a wide range of shapes, colours (but most often black) and sizes, they eat the eggs of root flies and also aphids that have fallen off their plants. Some of the more nimble species even climb plants to catch and eat aphids. Rove beetles, which resemble minature earwigs as they have short wing cases, behave in a similar way and climb plants in search of aphids. Many aphids are eaten by the maggots of hoverflies and are probably the most voracious aphid predators of all. The lace wing fly larvae also feed mainly on aphids. Adult assassin flies kill and eat small caterpillars and aphids. Predatory midge larvae are very small orange grubs that feed on aphids and may be found living within their colonies. Ichneumon flies and other small related insects act as parasitoids – that is, lethal internal parasites. Damsel bugs, which are often active at night, eat small caterpillars and aphids. Small, brownish flower bugs feed on aphids early in the season. It is also a little-known fact that as well as voraciously eating many tender plants, many common slugs eat many aphids but are not considered trustworthy as biological control agents!

In addition to this army of naturally occurring pests, many other organisms including many fungi, bacteria and other micro-organisms are effective in the biological control of pests. A large number of these have been tried commercially, but the only microbial product that is so far sufficiently reliable for garden use is *Bacillus thuringiensis*, a bacterium that infects – and different strains can kill a range of insects – but the one that is generally available is particularly effective against caterpillars. Commercial growers have used a fungus, *Verticillium lecanii*, to control aphids under glass. Other effective biological control agents for slugs, mushroom fly and vine weevil larvae are based on nematodes, but most commercially available biological control agents that are available for use in gardens or greenhouses are insects. Among these are the *Aphidius* wasps related to Ichneumons which control aphids, *Encarsia* wasps for the control of whitefly, *Amblyseius* and *Orius* bugs for the control of thrips, *Diglyphus* wasps for leafminers, *Phytoseilus* mites for controlling red spider mites, and *Abax* ground beetles for slug control.

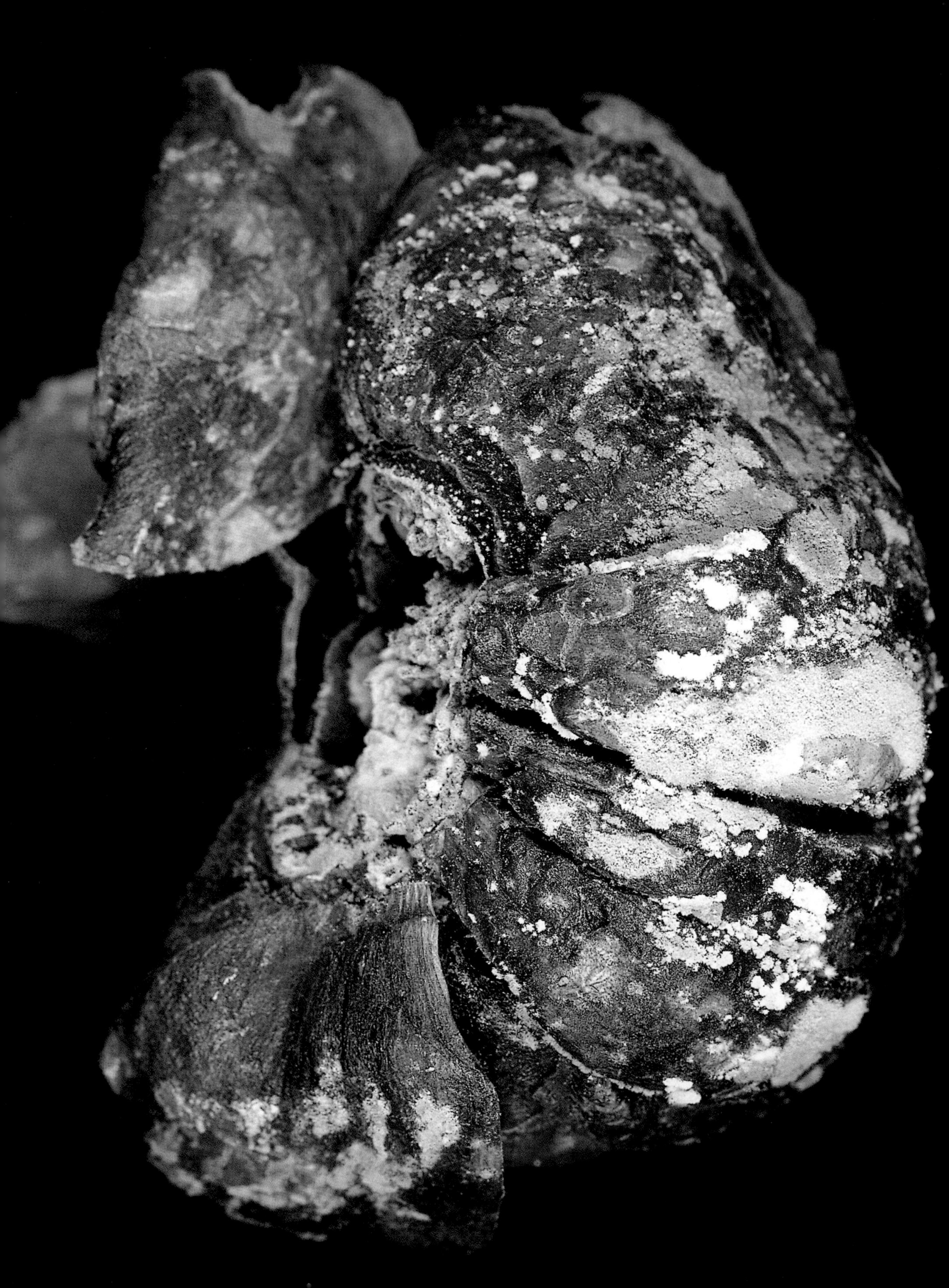

Recognising plant pests and diseases

i) Bulbs, corms and tubers

Bacterial rots – hyacinth yellows

Caused by *Xanthomonas campestris* pv. *hyacinthi* does not affect other bulbs. It invades the parts of the plant above ground first, then progresses down into the bulb. The bulbs rot either before or soon after planting. Often this happens in such epidemic proportions that no bulbs survive to emerge above ground or their blossoms are distorted. When a slightly affected bulb is cut across horizontally, a number of yellow spots will be seen in the scale leaves. If these scale leaves are squeezed a yellow slimey liquid oozes out. If observed under a microsope it can be seen that the ooze consists of numerous bacteria. A vertical section reveals that the yellow stripes or bands, continue down the leaf scales lengthwise and some may reach the basal plate of the bulb. These bacterial lesions connect with the foliage where they are confined to the vascular bundles. Those bacteria that reach the basal plate colonise it, then infect more scale leaves above it. Once this starts to happen the spots enlarge and connect with others on the leaves and scale leaves to form broad yellow bands. Ultimately other saprophytic bacteria invade, reducing the bulb to a decayed mass soon after the basal plate becomes destroyed. The roots are not affected but any offsets that are attached to the main bulb are usually infected. Bacteria from infected plants soon spreads by rainsplash, wind, insects and unwittingly on tools and humans to neighbouring plants which the bacteria enter through minute wounds in the leaves, flowers and flower stalks. These areas become water-soaked then yellow and finally blackish-brown. Hyacinths grown under luxuriant conditions are most at risk. The only way to avoid the disease is to buy plants from a reputable source that does not have the disease

Blue mould

Penicillium corymbiferum and other species cause damage to several bulbs including bulbous iris. It is seen when the outer scales are removed as a dark rot spreading either from the base or a lesion on the side of the bulb, these rots bear the blue mould. The infected bulbs are soft to the touch. Usually the rot does not extend to the outer fleshy scale, which becomes hard and chalky; however, occasionally the infection does spread gradually through the bulb, which becomes completely rotten. Infection penetrates up to the base of the bulb from the root ring. Often secondary organisms become involved in converting the bulb into a putrid mass. Bulbs that are less well infected often flower normally. Before purchasing bulbs look for

***Left*: Blue mould rot on a narcissus bulb.**

signs of infection. Nearly all bulbs offered for sale have been treated with a benzimidazole fungicide.

Dry rot of corms

Stromatinia gladioli occurs wherever gladiolus is grown. During the growth period six to eight-week-old plants show yellowed leaves which then become brown and dry as the plant dies. The dead plants show leaf sheaths that are blackened around the soil level. This area bears numerous black sclerotia. Beneath the leaf sheaths the corm is completely rotted away and frequently most of the cormels are affected. When buying corms it is essential to check for symptoms of dry rot.

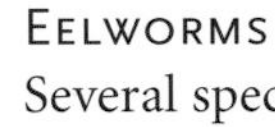

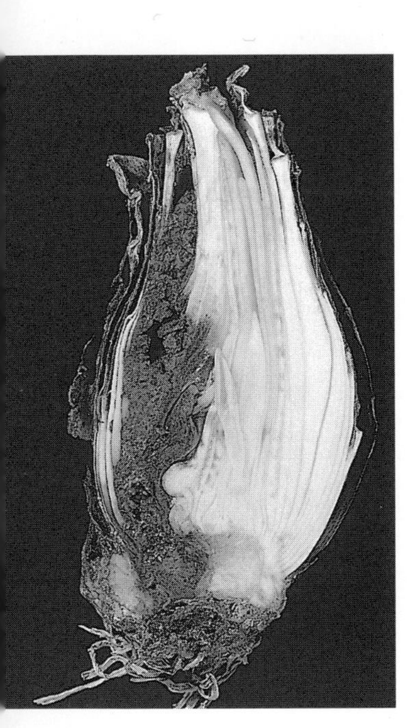

Eelworms

Several species of nematodes or eelworms are very destructive to bulbs, including the potato tuber eelworm (*Ditylenchus destructor*) and stem eelworm (*Ditylenchus dipsaci*). Neither is specific to bulbs and the potato tuber eelworm attacks a wide range of plants including potato, dahlia root tubers, lilac as well as many common weeds such as some mints and sowthistles, but is particularly damaging to iris. The stem eelworm feeds on hydrangea, oenothera, phlox and primula as well as narcissus, tulip and other bulbs.

The symptoms of the potato tuber eelworm vary but typically they cause a dark rot. In irises thin vertical black lines first appear in the scales and these coalesce and rot the bulb. The stem eelworms form both spongy and enlarged brownish areas which interfere with functioning of the tissues of the bulbs, even though the roots are rarely affected. At first these lesions stunt the plants, then they form a rot which spreads until the plant falls over. There are distinct races which occur on specific plant hosts which they can kill but they often leave some lightly infected plants which spread the stem eelworms through the soil to new hosts which they enter via natural openings or small wounds. Root lesion eelworms (*Pratylenchus* spp.) attack roots, tubers and rhizomes of several plants including clematis, delphinium, helleborus and thalictrum, as well as a number of bulbs and corms. The damage they cause is minor but admits bacteria which subsequently cause extensive rots. *Pratylenchus penetrans* causes root rot of narcissus, but is rarely transmitted to clean soil as it is generally killed when the bulbs and corms are dried.

Although bulb growers often dip bulbs in hot water to eradicate nematodes, this technique is rather too demanding for amateur gardeners.

Too high a temperature can damage the bulb and too low will have no effect. Nematodes are extremely difficult to eliminate from infested soil as they can exist in a dormant state for several years without feeding. They are difficult to control in soil, even with the few fumigants that professional growers, but not amateur growers, are permitted to use as soil sterilisers. Before planting bulbs in areas of the garden known to be affected by *Pratylenchus* spp., it is worth planting in some African and French marigolds (*Tagetes* spp.) since they are antagonistic to these nematodes. Although there is research into other methods of biological control with bacteria, this is not likely to benefit the amateur gardener for some time. At present the best advice for an amateur gardener with infested soil is to remove all weeds and not plant cultivated plants that are susceptible hosts in order to starve out the bacteria by denying them a food source. This is a long and imperfect process so when planting bulbs it is usually more sensible to choose another area of the garden known to be free of nematodes. Despite this precaution, great care is needed as several weeds and some cultivated hosts are symptomless carriers of infestation. Preventing the spread of eelworms is not easy but it is the only means of controlling eelworms available to amateur gardeners at present.

Fusarium basal rot of tulips

Basal rot is common and widespread on tulips in Britain. This is one of the most serious tulip diseases and often kills them. The disease is most rampant in hot summers after flowering has taken place, but is most apparent during the months of spring when conspicuous symptoms develop. Frequently the bulbs are killed in patches that can be easily confused with tulip fire (see p. 42) or the symptoms of drought stress.

The most characteristic early symptoms of infection on the leaves are wilting of the foliage and flowers. If bulbs are lifted and the bulb is split open, the basal plate and scale leaves are seen to be rotten.

Even apparently healthy bulbs can display symptoms such as gummosis later during storage. These are typically evident a few months after lifting. Softening and browning of the basal area is followed by a deep brown rot spreading through the inner bulb scales and often accompanied by an off-white mould. The causal pathogen, *Fusarium oxysporum*, is a common cosmopolitan soil-borne fungal pathogen which also attacks many plants usually causing foot rots, but the form that attacks tulips (f.sp. *tulipae*) is specific to them. Microconidia form in chains on short simple phialides. The spores are curved septate macroconidia which germinate rapidly to produce a mycelium of pale, cotton-wool-like hyphae. The fungus is often

***Left*: Stem eelworm damage on sectioned narcissus bulb.**

able to survive for a long time in the soil in the form of chlamydospores, a kind of resting spore, but even if these are not formed, the mycelium of *F. oysporum* f.sp. *tulipae* is often present in the bulb scales or survives as a saprophyte on plant debris. The spores – the macroconidia – form in prominent, light violet patches on infected bulbs.

The fungus spreads from infected soil into the bulbs through small wounds in the roots and is also present in the soil adhering to the bulbs when they are lifted and so may invade uninfected bulbs in store. Once soil is contaminated, it is practically impossible to eradicate completely so it is wise to plant tulip bulbs elsewhere. There are some differences in susceptibility between tulip cultivars, though none are highly resistant. Several fungicides have MAFF approval as dips to be applied to bulbs immediately after lifting, but only benomyl is available for use by gardeners as well as professionals. However, due to the selective survival of resistant strains of *Fusarium oxysporum* fsp. *tulipae* resistant to this (and other fungicides based on carbendazim), effectiveness has been observed to diminish following regular treatment.

Mites

Two mites commonly attack bulbs – these are the bulb mite (*Rhizoglyphus echinopus*) an acarid mite, and the bulb scale mite (*Steneotarsonemus laticeps*) which is a tarsonemid mite. Both occur on narcissus, but the bulb scale mite is a tiny, pale brown tocolourless mite which is much smaller than the bulb mite. Bulb scale mites are in fact so small that they are rarely detected unless their affect on plants is spotted. This appears as brown streaks from the base to the neck of the narcissus bulb, which is often smaller and softer than usual. If such bulbs are grown outdoors, unless the weather is unusually warm and humid, few symptoms are seen other than a reduction in flowering and the foliage dies earlier than normal. However, affected bulbs are unsuitable for forcing, because if they are used the increased temperature results in such a dramatic increase in their population that dust-like masses of mites can be seen on the necks of infested bulbs where they feed on the emerging foliage and flower buds. The foliage is at first bright green but later becomes malformed, streaked with yellow and may become scarred along the length of the leaves. If they flower at all, the infested buds produce distorted flowers which may be scarred. Hippeastrum is thus affected as well as narcissus. Here the first sign of infestation are red spots and streaks on the leaves, but later the leaves and stems become malformed. Under glasshouse conditions the life cycle takes only a fortnight so there are many generations throughout the year, but

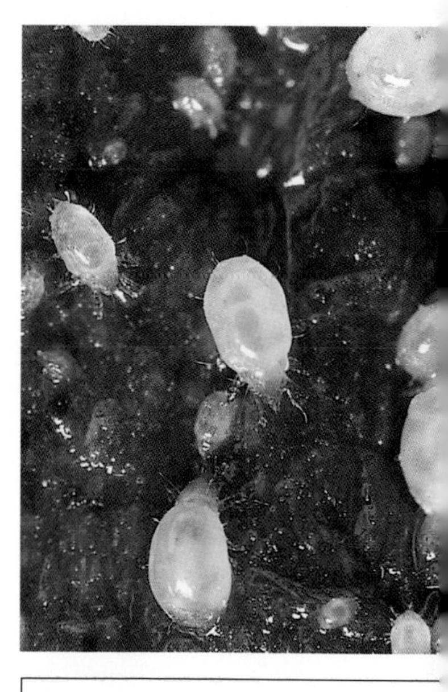

Left: ***Fusarium*** **basal rot on tulip.**
Above: **Bulb mites on narcissus bulb.**

their frequency is temperature-dependent. The mites remain in the bulbs when it is cold and are lifted up as the leaves start to grow. The best method of avoiding bulb scale mites is to ensure that any bulbs bought are free from the streak symptoms. Bulbs offered for sale may have been given a hot water treatment by the grower. This process is difficult for an amateur to carry out as it involves immersing the bulbs for either four hours at 43°C or for three hours at 44°C. For this reason any bulbs that show symptoms on lifting should either be destroyed or kept well away from healthy stocks as the mites are quite capable of walking to new hosts even in store rooms. Unfortunately, later infestations are very difficult to control with pesticides without damage.

The bulb mite (*Rhizoglyphus echinopus*) is a large yellowish-white mite frequently tinged with pink that can be seen with the naked eye, unlike the tiny bulb scale mite. The bulb is also commonly found living in the soil where narcissus bulbs are grown, but is also found on dahlia, freesia, gladiolus, hyacinth, lily, tulip and some other plants with bulbs, corms and tubers. In most of these cases it is likely that bulb mites can only attack bulbs that have previously been damaged, diseased or attacked by a pest. Once the mites have invaded the bulb, they reproduce quickly, rapidly extending the original wound, converting the bulb to a soft and rotten mass. In contrast, similar wounds on uninfested bulbs soon heal over. Mite-infested bulbs are often a source of fungal infections which can be spread by the mites in store to previously healthy bulbs. Reproduction is most rapid during the summer when the life cycle may take less than three weeks. There is an extra stage between the two nymphal stages and the adult stage, called the hypopus, which does not feed but attaches itself to insects and so is transported to fresh hosts some distance away. Some of the measures used to control bulb scale mites are also effective against bulb mites, for example, hot water treatment for one hour at 43.3°C and chemical treatment, but in general it is best to avoid the problem by buying only healthy bulbs.

Narcissus bulb fly

The larvae of the large narcissus fly (*Merodon equestris*) and the small narcissus flies (*Eumerus tuberculatus* and *E. strigatus*) are widespread not only wherever narcissus are found but also on a number of other hosts. Those of the large narcissus flies attack a number of bulbs, including those of hippeastrum, hyacinth, iris, leucojum, scilla, snowdrop and vallota as well as narcissus. As the dirty yellowish-white larva of the large narcissus fly is almost 20mm long when full grown, the damage caused when it tunnels into a bulb is initially quite serious, even though only one maggot is usually present, and is made worse as the tunnels fill with larval frass and decayed plant remains. The hole by which it enters can be found on the basal plate of the bulb. In contrast the greenish-white larvae of the small narcissus flies are only 8mm long and several occupy a single bulb. An immature larva of the large narcissus fly can also be distinguished from those of the small narcissus flies as it is fatter and has a more prominent dark brown, knob-like elongation on its hind end than the reddish-brown projection with whitish tubercles. The larvae of the large narcissus fly are not fully developed when the bulbs are lifted and usually the only sign of damage is the pin-sized entrance hole on the basal plate of the bulb. Any bulbs that are not detected and discarded at lifting are easily spotted when the bulbs are replanted a few months later, as they have softened, especially around the neck of the bulb. Infested bulbs that are not harvested or are missed during the inspection hardly ever flower and if they do produce foliage it is often yellowish, grassy and distorted. The smaller bulbs are generally more severely affected than larger ones. The adult large narcissus flies resemble bees as their bodies are covered in similar coloured hairs and they buzz as they fly. The shiny, black, adult, small narcissus flies are smaller with white marks on their abdomens. Although a few large narcissus flies can be found in February, these have emerged from forced bulbs and cannot survive; the majority are on the wing from the end of April to the end of June. The female usually lays a single egg either in the neck of the bulb or the soil nearby after crawling down the plant's shrivelled foliage on a sunny day, generally only about ten days after she has emerged from the pupa. The larva burrows down to the base plate of the bulb which it enters through a pin-sized hole which eventually enlarges to the diameter of a pencil when the larva emerges either there or through the neck to pupate under the soil surface for five to six weeks, before emerging as the adult. The adults of the small narcissus flies emerge from late April to early May to lay batches of five to 40 eggs on or around the bulbs. Most often the larvae enter the neck of the bulb and eat their way into the bulb, eventually destroying it. After pupating around the neck region for a couple of weeks the adults emerge to lay a second generation of eggs, some of which produce adults in the autumn, but these rarely lay eggs in turn. Many larvae overwinter in the soil and pupate in the spring.

The same methods are used for both the large and small narcissus flies, but for the latter prevention is more appropriate than cure as infested bulbs rarely recover fully. Even though hot water treatment at 44°C for an hour has been recommended to save those bulbs that are only slightly infested with the maggots of the large narcissus fly, this treatment is not encouraged for the amateur as it is not worthwhile on a small scale and there is a risk of physiological damage from the hot water. Instead of this measure or having an insecticide treatment, it is more sensible to eliminate the sources of infection. Any bulbs that have abnormal foliage should be lifted to find if they are infested by larvae, and burnt if any are found. Bulbs that are abnormally soft should also be destroyed. Some bulbs are pretreated with insecticide before being sold. It is not wise to dry off bulbs in the open air after lifting as this attracts the flies. Raking over the bulbs with soil, which is sometimes treated with insecticide, before the foliage dies back is a simple way of preventing the flies from being able to lay their eggs properly. This procedure can work even with naturalized bulbs to some extent, where it is the only precaution that can be taken.

Tulip fire

Tulip fire, caused by *Botrytis tulipae* and specific to tulips, can result in severe damage wherever these flowers are grown and has even been responsible for the total loss of tulip crops in some seasons. fire is first seen when a few plants with malformed leaves or shoots emerge above ground, scattered among the healthy ones. Frequently the whole shoot is reduced to a shrivelled, stunted shaft that is soon covered with conidiophores and sclerotia, especially in wet weather or humid conditions. Slightly infected leaves and stems may continue to grow and also produce conidia from the grey or greenish-grey diseased areas which enlarge rapidly to cover the whole leaf under moist conditions. Spread of the disease is hindered or completely checked by dry weather as the infected leaves often dry up, becoming split and torn. Some conidia are dispersed by air currents to the leaves, flower stalks or flowers of healthy plants where they give rise to leaf spots. Each yellowish spot is roughly circular, or somewhat elongated, slightly sunken and surrounded by a dark green, water-soaked area. These spots are readily seen when the leaf is held up to the light. Most spots remain minute and are incapable of forming conidia as the fungus rarely survives for long, but occasionally comparatively large spots or blotches merge and spores are produced. Nevertheless, tulips that are grown for cut flowers are spoilt by the small spots or blisters on the petals. flower buds that fail to open may become covered with mould. Dark brown patches that extend along the flower stalks sometimes bear black sclerotia. After the external papery scale has been removed, the outermost fleshy scale may reveal depressed circular lesions which also often bear sclerotia. Sometimes bulbs may become encased by masses of sclerotia after they have rotted. Newly planted tulip bulbs usually become infected through contamination of the parent bulb by conidia that are borne on its surface or mycelium that survives in the lesions on the outer scales. Although infection is often blamed on sclerotia in soil or on the bulbs, especially at the base of the old flower stalk, few sclerotia over two years old can still germinate. Inspect any bulbs that are to be purchased for sclerotia and reject any with blackened fleshy bulb scales. As a further safeguard, dip all bulbs before planting for 15 to 30 minutes in a suspension of a general garden fungicide that contains a benzimidazole fungicide, such as carbendazim, benomyl or thiophanate-methyl. This will also give some protection against several other important bulb diseases. Tulips should not be planted for at least three years on land where infected bulbs have grown and during this time spreading soil to clean beds should be avoided. The spread of disease to neighbouring plants is fast, so any infected tulips should be destroyed without delay. Spray the remaining healthy plants with a general garden fungicide that contains a dithiocarbamate.

***Left*: Tulip affected from tulip fire.**

(ii) Flowers

Earwigs

The common earwig, *Forticula auricularia*, is often the cause of holes eaten into the petals of a wide variety of flowers, and can be found hiding in them during the day and feeding on them at night. Earwigs congregate in other suitable crevices and inside dead plant material by day and also overwinter there. The female earwigs lays up to 100 eggs which she guards when they hatch in the spring with a second batch in the autumn. Although earwigs have wings they usually crawl to the flowers that they eat so damage is localised. Their diet includes aphids and other small insects as well as flowers and buds, but this asset is not usually enough to prevent their control if the damage they cause is sufficiently severe. Both the plants and the crevices where the earwigs hide should be treated with a suitable insecticide in the evening. It is also worthwhile tidying up the garden and glasshouse, although some gardeners purposely lay traps of cardboard, wood or sacking so that any earwigs that shelter there can be destroyed. Alternatively a trap can be constructed by filling a flowerpot with straw and then placing it upside down on a cane so that the earwigs can climb up to their doom. All of the methods of control depend on finding the earwigs, so if plants are moved they should be inspected, particularly those taken into glasshouses from the garden.

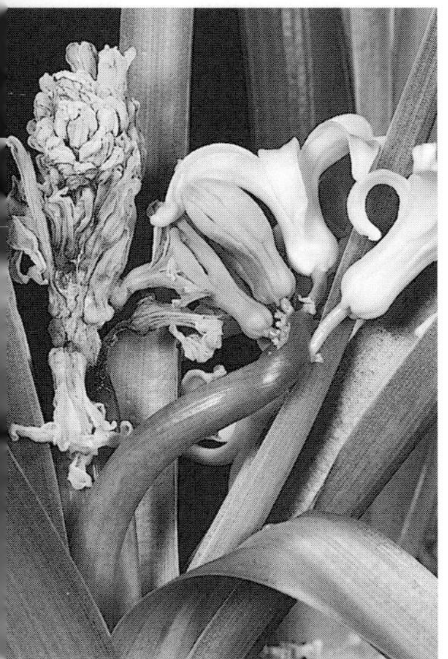

Grey mould (*Botrytis cinerea*)

In most countries, grey mould is probably the most common disease of vegetables, ornamentals, tubers, corms, bulbs, fruit, as well as many arable crops, and certainly is the most common disease in glasshouses. Predominantly a blossom blight and a fruit rot, it can also cause damping-off, stem canker, leaf spots and root rot. Infection often spreads to other adjacent produce after harvest.

Small yellow-tan, depressed leaf spots enlarge eventually coalescing to cover the leaf, then sporulate as a light grey felt on the dying tissue. Plants are rarely infected directly by conidia, but indirectly though decaying petals or leaves that stick to healthy leaves. After diseased fruit or fleshy stems develop soft and watery pale brown rots, grey mould sporulates profusely on them as they split open. Squat black sclerotia may form, submerged within the tissue as it eventually wrinkles and dries.

Seedling infection is worse in cold damp soil. Bulbs, corms and tubers are often infected while still in the ground. Lesions generally start at the crown or base. These appear soft and watery then enlarge, turn brownish spongy or corky, lose weight. Mycelium develops between decayed bulb scales, or on the surface of corms.

Botryotina fuckeliana is generally found as its asexual anamorph, *Botrytis cinerea*, which produces abundant long greyish, conidiophores with branches, bearing pale one-celled, ovoid conidia, resembling bunches of grapes.

Generally, *Botrytis cinerea* overwinters in the soil on plant debris as mycelium. Sclerotia or infected plant debris can be spread with seed. Growth and sporulation during cool damp weather favours release and germination of conidia, and establishment of infection through wounds, or via mycelium, on old flowers or dead foliage.

Remove infected debris both from the field and fruit stores. Reduce humidity in greenhouses and store rooms by providing suitable ventilation, so that plants can dry quickly in humid weather. Avoid harvesting fruit, such as apples and pears, and vegetables in damp weather. Store in a dry and cool environment. A few pre-harvest fungicide products are available to amateur gardeners, but chemical sprays are only partially successful under heavy disease pressure.

Some foot rots may also be caused by grey mould (*Botrytis cinerea*) that can grow down the plant inside or over the outside of the stems and between the leaves. On bedding plants, grey mould produces leaf spotting of various types which may later behave as a general decay organism if widespread necrosis occurs. Leaf spots of many sorts develop as a result of infection by a large number of seed and soil-borne mould fungi like *Alternaria* or *Phyllosticta.* Unspecialised bacteria, like *Pseudomonas* thrive in glasshouses and cold frames.

Avoid waterlogging, high humidity or any other influences or conditions liable to weaken or stress plants. Apply fungicide spray to plants after pricking out to prevent infection by mould.

Petal blights

Petal blights occur on number of herbaceous plants and shrubs. Chrysanthemum ray blight caused by *Didymella chrysanthemi* is another North American disease that has now become endemic in many parts of Europe, whereas chrysanthemum petal blight caused by *Itersonilia perplexans* has both a longer history and a wider host range, including anemones. The former primarily affects the shoots as well as the flowers. White and other light-coloured flowers develop reddish lesions which appear brown on darker flowers. Soon after they appear the flowers droop and rot. Although the disease can be spread by cuttings, it is more likely to occur through the pycnidia that form on dead plant debris. This debris

***Above Left*: Male earwig on leafshoots.**
***Left:* Grey mould infection of hyacinth.**

should be removed and burnt. If infection is anticipated appropriate fungicides can be used as foliar sprays or as a dip for cuttings prior to rooting. Petal blight affects the outer florets which appear glassy at first, then become shrivelled and brown. Some fungicides reduce petal blight if applied fortnightly. Rhododendron petal blight caused by *Ovulina azaleae* occasionally causes serious damage after rain in some areas of the northern range of rhododendron areas. Affected blooms are first spotted with small, light-coloured lesions. These merge together before the flowers droop and the invading fungus forms a whitish, sporulating mass. Sometimes sclerotia form, but in Europe spread is largely by conidia on the old diseased blooms which are dispersed by wind, rain splash and possibly insects. Infected flowers should be removed, whereas if infection is expected, apply an appropriate fungicide as the flower buds open. Lily disease is caused by grey mould, *Botrytis cinerea* which affects the blossoms on a wide variety of flowers including lily. Peony blight (Botrytis paeoniae) causes a bud blight in peony which resembles grey mould.

Bud blast of rhododendron and azalea

Pycnostysanus azaleae which causes bud blast is usually first noticeable in late autumn as a brown discoloration on some of the young flower buds of rhododendrons. These brown patches continue to develop and spread throughout the winter often becoming silvery grey. Although the bud becomes completely dry and eventually dies, it rarely drops off the flower stalk even though when the lateral buds are affected entire twigs may die. Leaves may suffer from scorch. In the spring following first infection, tiny, black bristly pinhead-like structures, up to 2mm high and 0.5mm wide, emerge from the affected buds, these are the coremia or synnemata which bear conidia. Fresh coremia continue to be produced from dead buds for three or more years. The disease is widespread in England wherever *Rhododendron ponticum* is common and so is particularly prevalent and especially severe throughout the sandy areas of Surrey and Berkshire. With bud blast and twig blight of azaleas and rhododendrons in Europe, North America and New Zealand, terminal flower buds are first affected; later lateral leaf buds and stems are attacked. When a twig is infected flowers fail to develop; leafy shoots become necrotic and may carry viable conidia for several years. It is easy to spot infected buds which become dwarfed, turn brown and later often silvery grey. It can sometimes be so serious in parts of southern England that the flowering of evergreen rhododendrons and azaleas appears very patchy or lopsided. Where these heavy attacks are common, bud blast is usually associated with large populations of the

Left: **Petal blight on chrysanthemum.**
Above: **Bud blast on rhododendron.**

leafhopper, *Graphocephala coccinea*, although their exact interrelationship has not yet been fully determined. It is possible that the damage caused to the buds by the leaf-hoppers as they lay eggs may provide wounds for the fungus to infect, since it appears unable to penetrate healthy tissues and also that the female leafhoppers act as a vector by carrying conidia to the buds. Although some species of rhododendron and hybrid varieties seem more resistant than others to bud blast attack, none is particularly satisfactory.

All affected buds and should be removed and destroyed by early spring. Spray with an approved fungicide, with a wetting agent, just before flowering and at monthly intervals thereafter when the attack is severe. If heavy attacks have been experienced, leafhoppers may be controlled by insecticide sprays at three weekly intervals from mid-June.

Thrips

Several species of thrips attack flowers. These include the gladiolus thrip (*Thrips simplex*), rose thrips (*Thrips fuscipennis*), carnation thrips (*Taeniothrips atratus*), glasshouse thrips (*Heliothrips haemorrhoidalis*) and onion thrips (*Thrips tabaci*). When the gladiolus thrips attack the flowers or foliage of gladiolus and related plants they mark them with tiny flecks as they feed. If the population is high enough this causes the flowers to wilt and die during late summer, often followed by attacks on the corms after they have been lifted, unless these are treated with a suitable insecticide. In the spring the female thrips start laying their eggs in plant material and continue through the summer, so populations soon build up. Rose thrips also fleck the flowers of roses and a number of other flowers. These brown thrips can be severe in glasshouses where they can overwinter in cracks in the structure of the building as well as other crevices. Their life history is similar to the gladiolus thrips. Carnation thrips also resemble the rose thrips as they attack plants outside as well as in glasshouses. Onion thrips are the ones most commonly found in glasshouses where they attack the flowers of many plants. Thrips often attack plants that are growing poorly under dry, warm conditions in the glasshouse or the garden. If these plants are kept well watered and cool damage can be limited. An appropriate insecticide can be applied as soon as the thrips or symptoms of their presence are seen.

Weevils

Both the apple blossom weevil (*Anthonomus pomorum*) and the leaf weevils (*Phyllobius* spp.) attack flowers. While the leaf weevils are rarely found damaging flowers, the apple blossom weevil is regularly responsible for the non-emergence of apple flowers in the spring. If such flower buds are broken apart they reveal stages of the weevil inside from larvae, pupae to the dark adults, this proves that they and not frost was responsible for the damage. Once the weevils have overwintered in crevices in the bark of apple trees and inside nearby debris they crawl up the tree to feed the young leaves. At the same time eggs are laid in the flower buds providing the larvae with a diet of stamens and other flower parts, thus preventing the buds from opening normally. The new generation of weevils emerge around midsummer but soon disappear into hibernation after having fed on leaves for a few weeks. If an ornamental crab apple is known to be heavily infested it may be worth spraying the leaves of small trees with a suitable insecticide.

Far Left: **Thrip damage to rose bloom.**
Left: **Silver-green leaf weevil.**

(iii) Leaves and buds

ADELGIDS

Adelgids can be distinguished from aphids, to which they are related, as they are smaller (1mm) and confined to conifers. Adelgids often migrate between two hosts, feeding on the needles and young shoots. When spruce is a host, the infestations are clearly defined as the tree's needle bases swell to form a characteristic gall that resembles a small distorted cone. However, these galls are not formed on other conifers even though the vigour of the trees lessens as the adelgids cause the shoots to deform and the needles to yellow and drop off. Adelgids also produce distinctive woolly patches of white wax. Even in large trees the presence of adelgids is generally all too obvious as they excrete honeydew that rains down on whatever is below which soon becomes stained by sooty moulds. Adelgids are found on a range of conifers. In addition to spruce, the Douglas fir adelgid (*Adelges cooleyi*) is found under the tree's needles which it changes to a mottled yellow colour, but although it forms galls on Sitka spruce it is not seriously damaging. The wool-producing generations of the spruce gall adelgids (*Adelges abietis* and *A. viridis*) occur on the needles and bark of larch, but the pineapple-like galls are found on both Norway and Sitka spruce. The galls of *A. viridis* are produced at the shoot apex and prevent further growth, whereas those of *A. abietis* are form at the base of the shoot and do not. However, *A. abietis* can complete its life cycle on spruce and so can be an important pest in Christmas tree plantations. There are some adelgids where spruce is never or rarely attacked and galls do not form. The hosts of these adelgids include silver firs, and Scots and Weymouth pine. Although it is possible to control spruce gall adelgids with insecticides providing sufficient wetting agent is included, this is rarely practical on large trees.

APHIDS

Aphids, also commonly known as greenfly or blackfly, are amongst the most familiar, damaging and frequent plant pests, different species of which attack virtually all garden plants, some on the foliage and others on the roots (see p. 113). Most common aphids are little more than 1.5 to 3mm long, with thin elongated legs and antennae, and have characteristic tubes called siphunculi or cornicles on their rear end which produce a waxy defensive fluid. The life cycles of many aphids are rather complex. At some times of the year aphids result from eggs laid as the consequence of mating, while others are borne by parthenogenesis without mating. Although several forms are produced within a species throughout the year, including

***Left*: Adelgids on needles of young scots pine.**

some with wings, most of these different forms resemble those forms of other aphids fairly closely, so different species of aphids are generally not easily distinguished by an amateur and in any case they are usually controlled by the same methods and chemicals.

Apart from the swarms of winged 'migrant' generation forms that areborne long distances on calm air currents in summer, most aphids move rather slowly and rarely crawl far from their colonies on their host plants. Nonetheless, in the autumn many aphids migrate to a winter host, usually a woody tree or shrub where they lay eggs that can survive in a crack or bud until spring before hatching into winged females. These winged females then give birth to repeated generations of female aphids by parthenogenesis. As the season progresses more and more of these aphids are born with wings and these migrate to the host plant which is favoured during the summer; usually this is a cultivated or wild annual or herbaceous plant. During the rest of the summer more winged and wingless generations of female aphids are produced until winged female aphids migrate back to the winter host to produce wingless female aphids. These wingless female aphids lay the overwintering eggs after mating with winged males. Some species of aphids remain on a herbaceous host plant all year round and do not have a winter host. In these species the males are usually wingless. In some species, sexual reproduction is unknown and all reproduction is parthenogenic; usually such species overwinter as nymphs or adults which may continue to reproduce if conditions are suitable. During the summer female aphids can produce as many as eight young every day for up to three weeks. In eight days these young will have matured and started to reproduce themselves, so in a very short time an insignificant infestation can become serious. As they build up in numbers, aphids can cause conspicuous distortions and leaf-fall on their summer hosts, both as a result of feeding on leaves and shoots by sucking up sap and also because of the toxicity of aphid saliva. Sometimes other symptoms result from the virus and other diseases that are also spread when the aphids feed or enter through the feeding scars. The sweet, sticky honeydew which is excreted by aphids blemishes the leaves especially when it is messily colonised by sooty moulds.

Aphids attract a wide range of natural predators and parasites. Among the more voracious are ladybird beetles, both adults and larvae, some bugs and the larvae of lacewings and hoverflies. Among the parasites which feed on aphids are the larvae of parasitic wasps which consume their bodies from the inside before leaving them (in the form of an adult wasp) through a hole in the empty swollen skin that remains attached to the foliage. The pupae of the gall midge (*Aphidioletes aphidomyza*) which lives entirely on an aphid diet are available by mail order. Once the package is received the contents of pupae mixed with vermiculite are tipped onto a square of damp newspaper covered by a clean plant pot, then left for a week for the gall midges to emerge. Where aphid infestation is already heavy, the gall midge is often applied seven days after a systemic aphicide that does not kill predators has been applied. If ants are present these should be controlled as they remove and kill the gall midge larvae, but insecticide should not be applied directly to the plants.

Among the most common and troublesome species are the beech aphid (*Phyllaphis fagi*) which forms a white woolly mass under beech leaves; the black aphid (*Aphis fabae*), a dark green aphid that attacks a wide range of herbaceous plants and overwinters on viburnum and the spindle bush; the cherry aphid (*Myzus cerasi*) that attacks cherry and various weeds; the chrysanthemum aphid (*Macrosiphoniella sanborni*), a dark shining aphid that distorts the flowers and spreads viruses; the fern aphid (*Idiopterus nephrelepidis*) which attacks glasshouse ferns, cyclamen and violets, sometimes outdoors in mild areas; the leaf-curling plum aphid (*Brachycaudus helichrysi)* that distorts the leaves of *Prunus* spp. and also occurs on a variety of garden plants which it stunts; the mottled arum aphid (*Aulacorthum circumflexum*) which attacks many glasshouse plants and outdoor periwinkles, often spreading viruses that cause colour breaking and mosaics; the peach-potato aphid (*Myzus persicae*) which is responsible for transmitting a number of viruses on a wide range of glasshouse and outdoor plants; the potato aphid (*Macrosiphum euphorbiae*), green or pink aphid that has a wide range of summer hosts but overwinters on rose; the rose aphid (*Macrosiphum rosae*) is the green or pink aphid most commonly found damaging the buds on roses, but it also infests holly, scabious and teasels during the summer; the spruce aphid (*Elatobium abietinum*), a green aphid that is well camouflaged between the needles until they yellow, brown or drop off; the tulip bulb aphid (*Dysaphis tulipae*) is found under the outer scales of several bulbs where it attacks the developing shoots; the water-lily aphid (*Rhopalosiphum nymphaeae*) is found on many aquatic plants during the summer but overwinters on species of *Prunus*; the woolly aphid (*Eriosoma lanigerum*) is found on apples and some other rosaceous shrubs, often on pruning wounds and grafts which swell and split as a result.

The best way to avoid the worst infestations by aphids is to prevent the build-up of large populations by inspecting plants, especially those about to be purchased, and treating early. On the woody hosts, aphids are often controlled by winter washes, but this does not completely control woolly aphids as they are difficult to wet. A number of aphicides have been developed for use on the summer hosts, but some of these also affect the predators and so are best avoided. However, it is generally recommended that some of the systemic aphicides that do not harm predators should not be applied during drought conditions unless the plants have been well watered first. Extra wetting agent is often recommended, especially for controlling woolly aphids. One way of controlling water-lily aphids is to submerge the infested plants so that fish and other aquatic animals can pick them off and eat them.

Beetles

Although there are many species of beetles, very few are common garden pests. Beetles differ from other insects as their two tough, scale-like front wings have become thick wing-cases or elytra, and have a protective function rather than being of use while flying. Beetles have biting mouthparts with well developed mandibles, but in the weevils these are at the tip of an elongated snout or proboscis that protrudes from the front of the head. The larvae of the various beetles are extremely diverse in their structure. Some larvae are covered in a burnished protective cuticle, whereas others resemble fat grubs with a soft fleshy skin. Although beetle larvae possess three pairs of thoracic legs, weevil larvae have none. Among the most troublesome beetle pests are the flea beetles which are common on vegetables but also attack a range of herbaceous border and rockery plants including alyssum, anemone, godetia, iris, nasturtium, stocks and wallflowers. These tiny beetles (2.5mm long) vary greatly in colour and some have stripy patterns, but many are predominately blue, purple or black and are able to jump actively from plant to plant. flea beetles damage plants, particularly young seedlings suffering from drought, by eating out tiny holes in the leaves. The adults and larvae of the lily beetle (*Lilioceris*) are found eating the leaves, stems and seed pods of a range of liliaceous plants and are especially damaging on lily, *Fritillaria* spp., *Nomocharis* spp. and *Polygonum* spp. The adults are conspicuous because of their bright scarlet bodies, whereas the larvae are reddish grubs that are typically covered in their own faeces. The water-lily beetle (*Galerucella nymphaeae*) is another leaf-eating pest. Both the dark brown adults and larvae can shred the leaves of the water-lilies on which they feed into tatters by devouring out long, irregular holes. The adults overwinter in the dead aquatic vegetation, becoming active in midsummer when they lay their eggs in clusters on the upper leaf surfaces of the water-lily pads. After feeding together the larvae separate and continue to feed until they pupate. Usually there are two generations each year. One way of destroying the larvae and adults is to hose them off the leaves or weigh them down under water so that any fish present can eat them. If there are no suitable fish, some insecticides are effective.

Birds

If a garden is reasonably near to woodland or hedgerows plenty of birds can be expected, especially in winter and spring. While most gardeners appreciate the presence of birds in the garden, there are a few birds that are not welcome because of the damage they cause, even though they may themselves be beautiful. The bullfinch is a good example of a highly attractive but extremely destructive bird pest. This stocky finch has a black cap to its head, grey wings with white and black bars; the male having a brilliant scarlet breast; the female is a somewhat more pallid version of the male. To someone with a wild garden the buds that bullfinches devour is a small price to pay for their beauty but anyone who wishes to see forsythia, lilac, almond, cherry, apple, viburnum and wisteria at their best would wish to scare them away – or worse – as soon as possible. Some other finches, such as chaffinches, greenfinches and hawfinches, can also be guilty of eating buds, as can house sparrows and the various tit species, but none of them can strip a tree as efficiently as a pair of bullfinches in early spring. Hawfinches often damage magnolia buds so badly that the flowers are mangled. Tits tend to attack the open flowers of camellia, red rhododendron and sweet pea in their search for nectar. House sparrows appear to mindlessly vandalise the petals of crocus, polyanthus, sweet-peas and violets which they tear into tatters and also eat newly sown grass seed. Shrubs and trees, such as cotoneaster, holly, pyracantha, rowan and stranvaesia, which are grown for their displays of autumn berries, are frequently raided by flocks of blackbirds, thrushes, jays, starlings and wood pigeons which soon strip them bare. Where there are shrubs with varieties that have berries of a somewhat unusual colour these are frequently avoided by the winged marauders. Divots are often dug in lawns by birds such as blackbirds, rooks and starlings while they are exploring with their beaks for the leatherjackets and chafer grubs which they eat in great quantities. Similar peck holes often appear if green woodpeckers are present in nearby woods as they will readily descend and spreadeagle themselves over ants' nests on lawns while they lick up the ants with their tongues after pecking away through the grass. Although this behaviour may be considered beneficial by those who dislike ants, this sort of damage to lawns is easily avoided by eradicating the insects. Starlings can be a nuisance as they occur in such large flocks that their droppings contaminate the trees where they roost and occasionally branches can be broken by the sheer weight of flocks or individual large birds. flocks of Canada geese which frequent lakes and ponds can also foul the surrounding grass with their green slimy droppings. The best way to avoid bird damage is to scare the birds away. There are several ways of doing this – some depend on attracting their sense of sight,

Left: **Lily beetle on damaged lily leaf.**
Above: **Pigeon damaged cabbage seedlings in organic vegetable garden.**

others rely on sound or touch. The traditional scarecrow is an example of the former method, but is only effective so long as it is moved about regularly. Some commercially produced scarecrows are capable of swinging in the wind. Strips of aluminium foil that glitter and twist, sometimes showing a colour on the reverse side as they are blown by the wind, can be effective without being too obtrusive, but can be bent to make a noise as well. In general noisy birdscarers such as those based on clappers, cartridges or gas-fired bangers are not appreciated by gardeners' neighbours. Netting cages around shrubs and fruit bushes and black thread or string hung with rags strung out between rows of plants effectively discourage birds from sweeping in to feed but can be rather unsightly. Newly sown grass seed can be protected by placing a covering of hay, straw, twigs or matting over it until it can establish. Sticky substances can be used to hinder perching on branches or the ledges of buildings but it is now illegal to use substances that glue the legs of small birds to twigs as with the traditional bird-lime used by the birdcatchers of old. Repellents that are sprayed onto plants have generally met with as little real success as those applied to deter cats and dogs despite several attempts to develop such a gardening aid. These days most birds are protected by law, though shooting or trapping can be carried out under some circumstances. These methods should be corroborated in each case by the appropriate authority such as the Ministry of Agriculture.

Black spot of roses

Diplocarpon rosae causes black spot of roses, which is restricted to the genus Rosaand is extremely common and widespread in most temperate and some tropical regions, especially under warm moist conditions. Rounded, dark brown to black blotches, often with irregular radially fringed margins which coalesce with other lesions, appear from late spring onwards on both leaf surfaces, but are most common on the upper side. floral parts and stems are also infected. Severe attacks cause abscission of yellowed leaves, often a further flush may also fall prematurely resulting in sickly shoots with dormant buds that give rise to feeble shoots in autumn which are susceptible to frosts, thus eventually killing or weakening the bush.

Diplocarpon rosae, has the conidial state *Marssonina rosae* which produces subcuticular acervuli on leaves and young shoots in summer, and from this black hyphae form a network with an internal intercellular mycelium with simple bulbous haustoria. Under moist conditions, elliptical hyaline conidia, constricted at the septum, are released from infected leaves then spread by rain splash or on hands, clothing and tools. Conidia lose viability rapidly, few surviving more than a month. The dark brown disc-shaped apothecia within fallen leaves contain inoperculate asci that forcibly eject eight oblong hyaline ascospores with two unequal cells, but if moisture is plentiful some apothecia produce conidia. Ascospores from apothecia found on decaying overwintered leaves in the spring throughout the north temperate zone enable new races of the fungus to arise and infect cultivars that previously were resistant. However, in Europe ascospores are superfluous for the survival of the pathogen as initial infections in spring are generally from conidia (mainly produced on overwintering lesions that develop on shoots in the autumn, with some from saprophytic mycelium in fallen leaves). A splash-dispersed subcuticular microconidial state (spermagonia) may also occur in dark spots on old decaying infected fallen leaves in the spring and autumn. The hyaline rod-shaped spermatia form terminally on sterigmata in a dense layer of spermatiophores. Two-celled conidia may also occasionally form in spermogonia. Germ tubes penetrate the cuticle directly, then spread through the subcuticular tissues, more commonly of the upper rather than the lower surface of the leaf, but very young leaves are not infected. In contact with free water or at a relative humidity of 85-100 percent conidia germinate over the range of 6-33°C (optimum 24°C), but below 18°C virulence decreases sharply. A week after infection lesions are apparent, with sporulation a few days later. Varieties resistant to powdery mildew and rust are of limited use as none is totally resistant to the numerous races of *D. rosae*. Partially resistant cultivars express symptoms that vary with regional environment and host conditions. Mixed planting of susceptible (often yellow flowered) varieties with more resistant (often red or pink cultivars) may lessen black spot. Severe spring pruning and destruction of infected leaves and shoots early in the season can delay disease. Improved drainage, avoidance of shading and of excess nitrate in fertilisers may also be beneficial. Some areas affected by air pollution are free from black spot as conidial germination may be inhibited by low concentrations of sulphur dioxide. Several fungicides are approved in the UK and should be used as instructed. In Britain the application of most protectant fungicides is timed to prevent inoculum in August leading to rapid spread in September. Several other common but minor leaf spots (including those caused by *Elsinoë rosarum* and *Sphaerulina rehmiana*) may be treated in the same way as black spot.

Capsids

There are three main species of capsids – the common green capsid (*Lygocoris pabulinus*), the potato capsid (*Calocoris norvegicus*) and the tarnished plant bug or bishop bug (*Lygnus rugulipennis*). Capsids are fast

running bugs that often drop to the ground once they are disturbed in foliage. They can damage flower buds, as well as young leaves and shoots, which they puncture, leaving a small brown spot with a hole in the centre which can tear or distort as the flower or leaf opens. flowers may fail to develop or become malformed if the growing point is damaged. The common green capsid is found on many flowering trees and shrubs, including fruit trees, forsythia, hydrangea and rose, as well as many herbaceous plants like chrysanthemum, dahlia, fuchsia, salvia, sunflower and a variety of weeds. The eggs overwinter in the fleshy shoots of a woody host. After passing through three nymph instars the adults lay the eggs during the summer for the next generation which become adult in late summer to autumn. These lay the eggs that overwinter. The potato capsid closely resembles the common green capsid, but is generally a problem on composite plants like chrysanthemum. The tarnished plant bug, which feeds on herbaceous and annual plants including weeds, is usually pale green with reddish-brown markings, but can be more variable in colour with green or brown nymphs. The adults, which overwinter inside hollow plant stems and in decaying and rank vegetation, emerge in the spring and lay their eggs in the stems and unopened flower buds of suitable host plants, often weeds such as dock, groundsel, nettle and sorrel. By late summer the adults from these wild hosts invade gardens to attack asters, arctotis, chrysanthemum, dahlia, nasturtium, poppy and zinnia. In order to prevent serious damage it is essential to inspect plants closely for the first signs of puncturing as several insecticides give adequate control if applied early enough. As capsid bugs fall off readily, the ground beneath the plants should also be treated. Weeds, decaying and rank vegetation should be tidied up as well because these provide food and hiding places for capsids.

Caterpillars

Caterpillars are usually the larvae of moths and butterflies, but they can be mistaken for those of sawflies (see p. 87). Many caterpillars feed on wild plants in small numbers but others are important pests. Caterpillars feed on their hosts plants with their strong-toothed mandibles, yet the adults are not harmful as they feed on liquids, particularly nectar, sucked up through a tubular proboscis. Most moths and butterflies are so well-known that they need little description, although many are quite bizarre. Nearly all have two pairs of wings covered by flattened scales which bestow the colours we see. The adults lay clusters of eggs, usually yellow or green, on the undersides of the leaves of a food plant. The larvae of many are also familiar as either hairy or smooth, fleshy caterpillars. Sometimes the caterpillars are

***Left*: Buff tip moth caterpillars.**
***Above*: Tarnished plant bug.**

conspicuously coloured, but many are so well camouflaged that they are often detected by the pellets of frass (excrement) that drop from and on affected plants. Most move around on three pairs of legs on the thorax and five pairs of prolegs, but the looper caterpillars have only two pairs of prolegs and propel themselves by forming a loop then stretching their entire body forwards. Another characteristic of caterpillars is the ability to produce silk which is often used to protect the larvae as well as the pupae, by the construction of a tent. The pupae are usually hidden in sheltered places among plants, in crevices or in the soil. Other larvae and pupae are protected inside the leaves which they have mined.

Although most caterpillars feed on foliage, others are stem borers or live in the soil. Many of the foliage feeders are defoliators which feed voraciously on leaves, making holes and ragged edges until often only the tattered leaf skeletons are left. Among these some are leaf-tying species which feed between or inside leaves which they have fastened with silk. Some of these are webbers, colonies of which live inside a communal tent. Among the defoliators are a number of very damaging caterpillars. One of the worst is that of the angle shades moth (*Phlogophora meticulosa*), a moth that has pinkish-brown wings with a central triangular band and marginal line of olive green. This caterpillar is somewhat variable in colour, from brown to dull or bright green, usually the latter predominates – all are dotted with white, with pale lines along the sides and back. Two or three generations can be found on a great variety of ornamental outdoor and glasshouse plants and many weed species, but are especially damaging to the buds and flowers of gladiolus and iris during late summer. The caterpillars pupate in the soil in cocoons built from silk and soil. The cream to pale brown caterpillars of the brown china mark moth (*Nymphula nymphaeata*) can live under water, protruding from inside a case made from two pieces of leaf which contains its air supply, although they can absorb oxygen from the water when young. The case is renewed as the caterpillar grows and while it hibernates the case protects it and then the pupa which forms inside. They feed on many aquatic plants, stripping off the surface of water-lily leaves, causing them to decay. The moths are rather fragile, with wings that are mottled irregularly white and brown or orange-yellow. Colonies of the yellow short-haired caterpillars of the buff-tip moth (*Phalera bucephala*) can strip the leaves from a range of trees and bushes, including cherry, rose and viburnum, if they are not destroyed in time. As well as causing severe damage to cabbages, the greenish caterpillars of the familiar black and white cabbage white butterflies (*Pieris* spp.) attack a number of ornamental plants including mignonette, nasturtium and stocks. The tomato moth (*Lacanobia oleracea*) can be a major pest of glasshouse carnations and chrysanthemums. The multicoloured, yellow tufted caterpillar of the vapourer moth (*Orgyia antiqua*) feeds on many trees and shrubs. There are three species of winter moths whose looper caterpillars defoliate many trees and shrubs and may injure their buds and flowers. These are the winter moth (*Operophtera brumata*), the March moth (*Alsophila aescularia*) and the mottled umber moth (*Erannis defoliaria*). As well as these there are several other defoliating caterpillars which are occasionally troublesome, but these can be controlled in the same way by insecticides or by biological control agents based on the spores of the bacterium, *Bacillus thuringiensis*, often referred to as BT. The powder containing microscopic spores of *B. thuringiensis* are usually supplied in sachets can be stored for up to six months if kept refrigerated below 10°C, dry and unopened. When required this powder is suspended in water and sprayed evenly but, in much the same way as a pesticide, not so that it can run off onto both sides of the leaves of the plants where caterpillars are actively feeding. If there is a severe infestation an appropriate chemical insecticide is sometimes applied first. It is best to apply *B. thuringiensis* as soon as caterpillar damage is noticed and it is important to make repeat applications if the plants are watered, there has been rain or further eggs have been laid. Within a few hours of consuming treated foliage the caterpillars stop feeding and die two to five days later. Younger caterpillars are the most susceptible. Pupae, adults and mature caterpillars that have stopped feeding in order to pupate are not affected. Other effective methods of control include picking off individual caterpillars by hand and crushing them. The female winter moths are wingless and can be discouraged by painting a band of grease around the trunks of susceptible trees. Several caterpillars belong to species that tie the leaves of their host plants with silk. In addition to a number of other tortrix moths which injure a wide range of outdoor and glasshouse plants, the caterpillar of the carnation tortrix moth (*Cacoecimorpha pronubana*) is common outdoors on a number of shrubs but is very damaging to a range of glasshouse plants as it feeds by tunnelling into their buds and bundling leaves together. As they drop from infested plants on a thread of silk if disturbed, they can easily be spread by clinging on the clothing of visitors. The caterpillars of the delphinium moth (*Polychrysia moneta*) which are initially brown, becoming green striped with white as they age, feed on the flowers, buds, seed capsules and leaves of delphinium, larkspur and aconitum after tying them with silk. Tortrix moth caterpillars are difficult to control on many ornamental plants as they remain hidden inside their bundles of leaves, so they are often most effectively controlled by being picked off by hand and burnt. However, if an attack is anticipated insecticide sprays can be an effective deterrent.

Some caterpillars produce protective webs of silk, including the hawthorn webber moth (*Scythropia crategella*), colonies of which colonise cotoneaster as well as hawthorn, the juniper webber moth (*Dichromeris marginella*) found on juniper, the lackey moth (*Malacosoma neustria*) which sporadically becomes a pest on a range of trees and shrubs and the small ermine moths (*Yponomeutra spp.*) that form their tents on a wide range of trees and shrubs. Webbers are best controlled by applying an insecticide as soon as the caterpillars are seen. If not, the webs can be pruned out or burnt off with a blowtorch. Tar oil is often used to destroy the overwintering caterpillars of small ermine moths but is ineffective against the egg-bands of the lackey moth which should be destroyed by pruning out and burning.

Crab apple scab

This ugly disease is common in most regions, especially in sheltered gardens. You can find the round black scabs on leaves and young shoots in summer. The fungus forms a network under the leaf surface, mostly on the upper side. Fruits and stems are also infected. If you do not treat scab it may eventually weaken the tree, as well as badly distorting the fruits. The scabs produce two types of spores. In summer, the conidia develop under moist conditions. These spores are released and spread by rain splash or on your hands, clothing and tools. Few conidia survive more than a month. The second type of spore, the ascospores, are formed within fallen leaves. However, ascospores are not the only way that the disease survives the winter. The first infections in spring generally emanate from conidia from scabs on older shoots or fallen leaves. Several fungicides are approved in the UK and should be used as instructed on the label. The main symptom is the dark brown-green, rounded blistered patches of scabs that form on leaves, twigs, sepals of flowers and fruits of apple (*Malus*). The leaves on the flowering spurs are particularly affected. Older infections become darker and more crinkled and may drop off prematurely. In spring the blistered, infected tissue on the twigs sporulates. Scab lesions may girdle and kill some smaller twigs, resulting in some crown dieback. Small dark green scabs on the fruit enlarge and crack as they grow and may continue to enlarge after harvest but are superficial – no rotting occurs. The causal organism on apples is *Venturia inaequalis* which also causes scab on a range of other *Malus* species. *Venturia pirina* infects pear. Both cause important losses of fruit quantity and quality. Many strains occur. Two types of spores are produced: conidia as well as the uniseptate ascospores. The fungus overwinters as an immature perithecia in ascomata which form deep within infected fallen leaves as the result of sexual reproduction between antheridia and ascogonia derived from different lesions. Damp weather is essential for ascospore discharge from the perithecia in spring. Heavy ascospore discharge at blossom time results in increased infection. Ascospores chiefly infect young leaves which rapidly form lesions, one or two weeks later producing velvety masses of short, erect, brown conidiophores bearing the bullet-shaped conidia that emerge through the epidermis. These conidia are blown or rain-splashed to other young leaves and developing fruit. Following infection the dark hyphae from the sub-cuticular mycelium conspicuously spreads across the leaf. Apple twigs produce conidia from lesions for one year only (unlike pear scab where conidia are produced on the same wood for several years, which is why pear scab is continuously so severe). Because of localised secondary spread there is often one major and several minor scab lesions on each fruit or leaf, but these may coalesce.

Fallen leaves should be collected and burnt. Prune out and burn scabbed young twigs from infected trees.

Several fungicides are currently approved only for professionals, though a few products are also available for garden use. Pesticides must only be used as directed on the label. Leaves are infected, as they open in the early spring, when rain wetness persists on their surfaces. It is then that chemical control is essential (almost always in late April or May) when copious ascospore discharge occurs from dead leaves left on the ground, and conidia washed from twig lesions are numerous. Once visible lesions can be seen on leaves and tiny fruitlets, little can save the crop from scabbing. Thus it is that severe scab on apples (and pears) at harvest is associated with short, wet and warm periods in early spring. Fully expanded leaves and fruits larger than 2cm in size are resistant to scab infection.

Left: **Scab on apple leaf.**
Above: **Fireblight**

Fireblight

The name of this disease caused by the bacterium, *Erwinia amylovora* is descriptive of the symptoms that it causes on apples, pears, amelanchier, chaenomeles, cotoneaster, pyracantha, stransvaesia, hawthorn and rowan or mountain ash, as well as other sorbus species. On most of these species the blackened twigs, flowers and foliage of the affected plants look as though they have been burnt by fire. This is another example of a disease originating in North America which spread to Britain and then on to most other European countries. In some countries it has become endemic, as it has in most of Britain, where it is now only a notifiable disease in the vicinity of apple and pear orchards, but there has been an attempt to eradicate it in Norway. The symptoms are sufficiently clear on most hosts. Even vigorous hosts can be transformed by a fireblight strike. Initially the dead flowers and leaves hang down on the twigs that have been affected. If these are not pruned away the infection can spread down the fruit spurs to the twigs from them to the branch and from the branch to the trunk. On these woody areas the bacteria form cankers in which they overwinter. In the spring the bacteria ooze out of the cankers onto twigs and these are attractive to bees and other insects which spread the slimy bacteria to flowers on new hosts. Later in the year bacterial slime can be spread to new hosts by rainsplash during summer storms. In this case the twigs are usually infected directly. Outbreaks of fireblight should be dealt with as soon as possible. Although some antibiotics are allowed in the USA, no effective bactericides are available within the European Union because of the fear of resistance to antibiotics spreading to human pathogens. Affected branches should be cut off and burned. In an area with numerous infected hawthorns in the hedgerows it is wise to consider planting shrubs and trees that are not affected by fireblight.

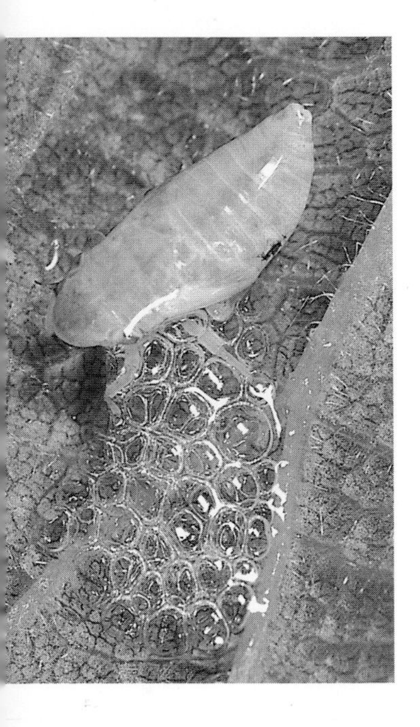

Froghoppers

The common froghopper (*Philaenus spumarius*) is the only species that is damaging to ornamental plants. The adult is an active, stocky insect about 6mm long which jumps long distances with its powerful hind legs and is usually seen in July and August. Adult coloration is variable from light yellow to dark brown with yellow markings, and nymphs are light yellow or green but are usually covered by foam. This foam or cuckoo-spit prevents the nymphs from drying out but can upset some gardeners by its appearance on favourite plants such as lavender, campanula, coreopsis, geum, lychnis phlox, rose, rudbeckia and solidago between May and July. As well as these cultivated plants, the adults lay their eggs on a wide range of wild plants during late summer, where they overwinter. The nymphs, which hatch out the following May, suck the sap of the host plants, causing them to wilt or become distorted. The nymphs are susceptible to several garden insecticides and can be washed off the leaves.

Gall midge flies

Although they vary in shape and size, all flies possess only one pair of wings; the other pair have been modified into protuberances called halteres which enable the insect to balance when it is flying. flies feed on liquids such as honeydew, nectar or decomposed organic material by sucking them up through specially evolved mouth parts. The larvae of most flies that feed on plants are maggots that have no prominent head or legs. Many of these maggots have mouth hooks by which they tear and rasp off pieces of plant material. The two types of fly that attack plant foliage are the gall midges and the leaf miners which are dealt with elsewhere. The gall midges are so small that they are not often noticed. The damage is done by the tiny maggots which have rather diminutive heads. There are several important species of gall midges, including the chrysanthemum gall midge (*Diarthronomyia chrysanthemi*) and the violet leaf midge (*Dasyneura affinis*). When the chrysanthemum gall midge was first introduced from North America it was much more damaging than it is now. The translucent whitish to orange-coloured maggots feed in 2.5mm long, lopsided galls that protrude from the upper surface of the leaves in light attacks but which also appear on the lower leaf surfaces, stems, buds and calyces in the more severe infestations that generally occur more frequently under glass. As a result of the more extensive attacks, the leaves, stems and flowers are distorted. Often the plants and cuttings taken from them fail to grow properly and the flower buds remain closed. Under glasshouse conditions chrysanthemum gall midges can produce up to eight generations each year, with more in the summer than the winter. The chrysanthemum gall midges lay their eggs in the buds, unfolded young leaves or the tips of the shoots, where the maggots burrow inside after they hatch. The larvae feed inside the galls, which become visible after two to four weeks in summer or over 12 weeks in winter, and then pupate in them, with the new generation of midges emerging after less than a month in summer (five months in winter). The violet leaf midge can be very damaging to cultivated and wild violets. The larvae are similar to those of the chrysanthemum gall midges but the symptoms that they cause are very different, as there is no discrete gall; instead the leaf edges roll upwards over the upper surface of the leaf. If the infestation becomes severe the leaves swell and the leaf rolls up so much that it cannot function and as a result the plant fails to grow or produce flowers. Occasionally the sepals and petals of the flowers are infested and become swollen and curl up. Outdoors there are four generations of midges each year which produce peaks of adult emergence in early May, mid-July, the end of August/early September and late October/early November. An extra generation is often produced under glass. Almost as soon as the adults emerge they mate and start to lay their eggs in the unfolded leaves, which never open properly as a result and stay curled over the larvae until they pupate there inside a cocoon. Prevention is much more effective than cure, so it is wise to inspect all chrysanthemum plants carefully for galls and check violets for curled leaves before purchase. It is sensible to quarantine newly acquired chrysanthemum and violet plants well away from other plants. All heavily galled leaves and stems in chrysanthemum and curled

***Left*: Froghopper nymph on leaf with 'cuckoo spit'.**

leaves in violets should be removed and burnt before the plants are treated with insecticide since it is difficult to achieve much success as the chemicals do not penetrate living matter easily.

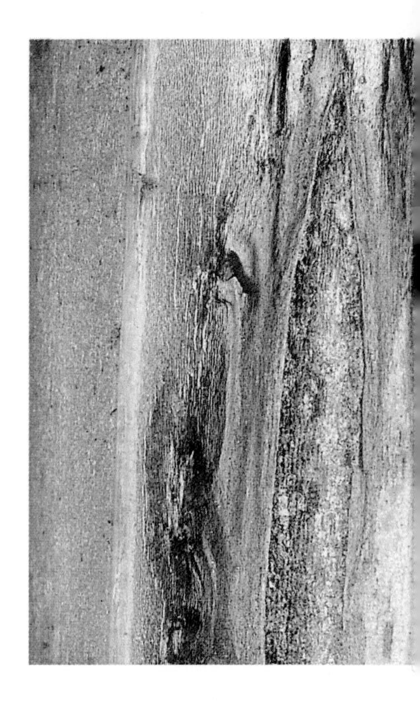

Leafcurls

Many ornamental almonds are affected by this disease which also affects peaches, nectarines and occasionally apricots that are grown outdoors in cool wet sites. After the buds burst in the spring the new leaves of infected plants are seen to be coloured bright red and yellow, and are much thicker than usual. As the season progresses these leaves become more distorted as they curl and twist with the infected areas bulging out and appearing to be covered with a white bloom. After this the leaves become brown and many drop off. This reduces the vigour of the tree so that fewer blossoms and fruits are formed. The pathogen involved is *Taphrina deformans.* Another related species *T. pruni* occurs on plums where the fruit are deformed into an elongated pocket-like structure which eventually become lighter in colour. There are also leaf curls on cherries, but often aphids cause similar symptoms so it is essential to look for the presence of aphids. In all these cases for most of the year the fungus that causes the symptom in the spring exists as a harmless yeast-like form that survives on the bark of the tree and overwinters there and in the bud scales. This yeast-like form enters the buds to infect the developing leaves in the spring. These infected curled leaves then produce ascospores which germinate to form the yeast-like stage. In the garden it is sensible to remove all the curled leaves and burn them, but it is not possible to control the disease at this stage. Instead, it is customary to spray the trees with Bordeaux mixture after leaf fall in the autumn and before the buds open in early spring.

Leafhoppers

Adult leaf hoppers resemble froghoppers as they jump around in a rather similar way but are much smaller (3mm), more delicate and tend to fly rather rapidly when attempting to locate fresh hosts. Without the protection of the foam that covers the nymphs of froghoppers, the initially inactive leafhopper nymphs, like the more active adults, also suck the sap, usually from the underside of a leaf. The points where they feed induce yellowing of the corresponding areas on the upper surface of the leaf. This mottling starts as several distinct, light yellow spots which coalesce into continuous pale areas. The areas where the nymphs have fed are usually covered by their moulted skins, sometimes called 'ghost flies', each of which

***Left*: Rhododendron leaf hopper.**
***Above*: Gall midge galls on stem.**

remains securely attached by its proboscis which the nymph had impaled into the tissues of the leaf before it moulted into an adult. Several leaf hoppers are troublesome in the garden and glasshouse. Despite its name, the glasshouse leafhopper (*Zyginia pallidifrons*) which is pale yellow with two dark chevrons on its back, attacks many outdoor ornamental plants and weeds like chickweed as well as stunting many indoor plants and killing their seedlings including calceolaria, chrysanthemum, fuchsia, geranium, heliotrope, pelargonium, primula, salvia and verbena. The rose leafhopper (*Typhlocyba rosae*) is somewhat larger than the glasshouse leafhopper and entirely yellow. It attacks roses, particularly climbing roses, causing stunting and premature leaf fall, especially during droughts. The rhododendron leafhopper (*Graphocephala coccinea*) was introduced in the 1930s into southern England where it is now well established. It is over three times as large as the rose leafhopper, with a yellow head, blue-green thorax marked with yellow, red and blue-green wings with scarlet stripes. The nymphs are yellow-green. Unlike the other two species, the rhododendron hopper does not appear to harm rhododendron bushes directly even when they are infested with numerous hoppers, but it has the reputation of vectoring *Pycnostysanus azaleae*, the fungus which causes rhododendron bud blast. This disease can be very damaging to many *Rhododendron* spp. but is generally reduced when rhododendron hoppers are controlled.

The life cycle of these leafhoppers varies. The glasshouse leafhopper breeds continuously under protected cultivation, taking from four to 12 weeks depending on temperature and other conditions during the year. The eggs are laid in the secondary veins on the undersides of leaves. The rose leafhopper has two generations each year from eggs laid just under the surface of young stem shoots. The first batch is laid in the autumn and hatches into nymphs in May and these develop into adults by June. They then lay eggs in July which hatch into nymphs in August/September. This second generation is usually of less consequence than the first. When the eggs of the rhododendron hopper are laid from late summer to autumn, they are inserted into slits in the flower bud scales where they remain over winter before hatching into nymphs in late spring and developing into adults by July. It is likely that the spores of *Pycnostysanus azaleae* enter the buds through these surface wounds.

Several insecticides can be applied as fortnightly sprays to the undersides of the leaves immediately the infestation is detected, or from mid-May when roses are likely to be affected. Some insecticides formulated as smokes can be applied in glasshouses. Best results are usually obtained following the application of insecticides during overcast weather when the leafhoppers are least active. Care should be taken to remove all the weed hosts of the glasshouse leafhopper, such as chickweed. Rhododendron hoppers are usually only controlled if budblast is damaging.

Leafminers

Most insects that mine into leaves are larvae, either the maggots of flies or caterpillars of moths, whose activities result in linear and blotch mines. However, the communal larvae of a geum sawfly (*Metallus gei*) form complex blotch-type leaf mines which virtually cover the leaf surface. Linear mines are caused by a larva that had meandered erratically through the leaf, so that thc outlinc of the mine that it excavated increased in width as the larva grew. In contrast, blotch mines are caused by one or several larvae excavating a blister-like cavity in the leaf which may become necrotic and if it is sufficiently extensive the leaf may be destroyed. There are several important leaf miner caterpillars: the azalea leaf miner (Caloptilia azaleella), the lilac leaf miner (*Caloptilia syringella*), the laburnum leaf miner (*Leucoptera laburnella*) and the rose leaf miner (*Stigmella anomalella*). The adult of all of these is a tiny moth. Those of the azalea leaf and lilac leaf miner have wingspans of about 13mm, whereas that of the rose leaf miner is less than half this, with the laburnum leaf miner intermediate in size. Azalea leaf miners cause blister-like mines on leaves, which often shrivel and drop, disfiguring azaleas grown indoors, as well as many of those outdoors in southeast England. The caterpillar is initially colourless but later turns green when it emerges and either fastens the leaf in a curl and proceeds to feed inside its protective cover, or on some varieties of azalea it excavates a new mine, often in a new leaf. The cycle can be completed in two months. Epidemics of laburnum leaf miner are frequent in England where many trees are badly affected, especially the younger ones whose growth may be temporarily inhibited. The mines start off linear but become a blotch that covers almost all of the leaf which may then turn brown and wither. The caterpillars fall from the affected leaves and pupate on the branches or the leaf litter on the ground below. The lilac leaf miner scorches the leaves and suppresses the growth of lilac and privet by excavating asymmetric blisters, often containing several caterpillars. Later the caterpillars emerge and hide in rolled-up leaves. There are two generations each year of this and the rose leaf miner. The rose leaf miner is rarely as serious, even though the linear mines are unsightly. All leaf miners are difficult to eradicate once they are protected in their mines but if applied early enough several insecticide sprays can give some control. Adults can be controlled by smoke formulations in glasshouses. If only a few leaves are affected, they can be collected and burnt.

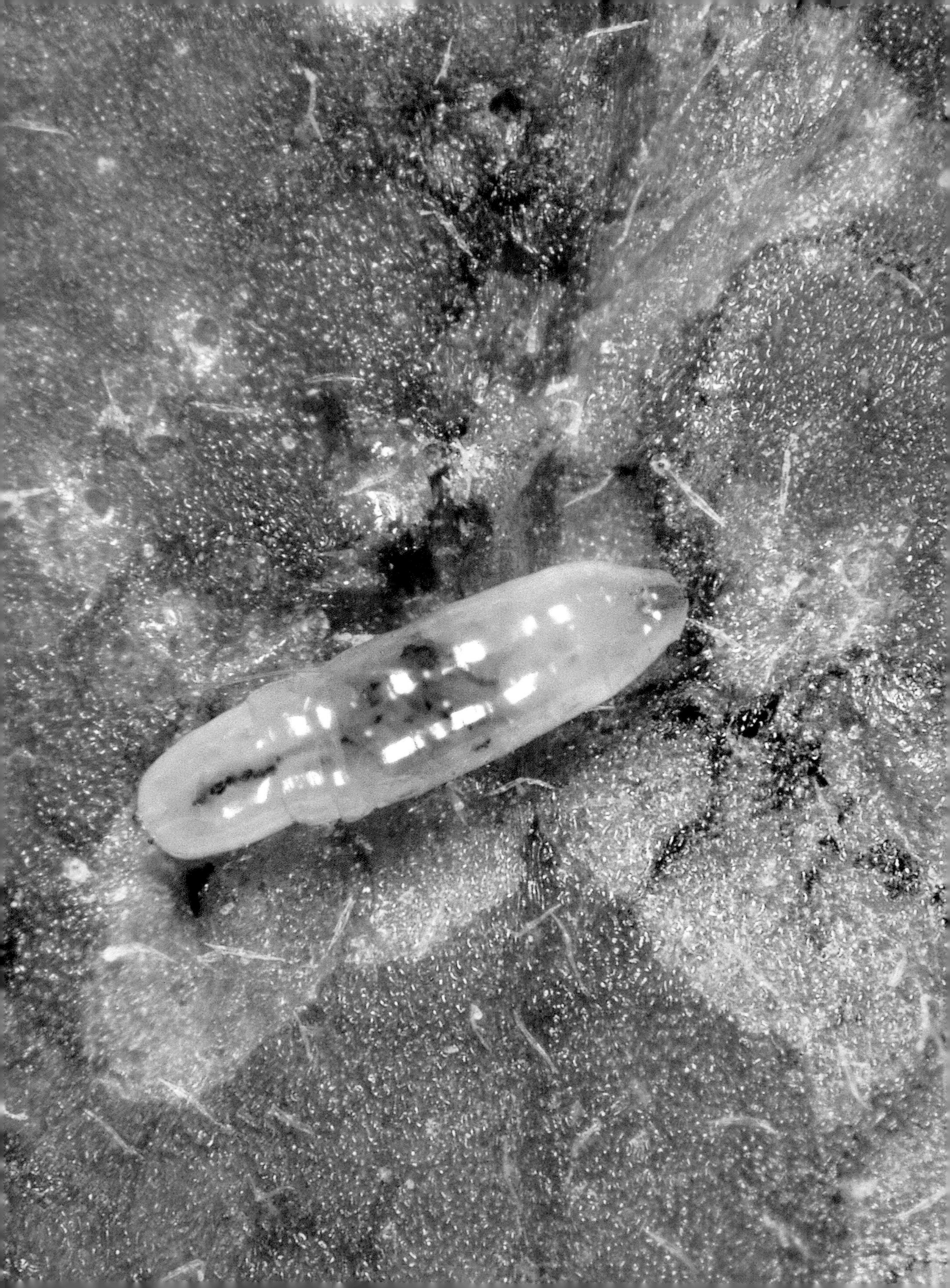

Some fly larvae cause linear and blotch mines on the leaves of a very wide range of herbaceous, climbers and woody plants. Among these are the carnation fly (*Delia brunnescens*), the chrysanthemum blotch miner (*Euribia zoe*), the chrysanthemum leaf miner (*Phytomyza syngensiae*) and the holly leaf miner (*Phytomyza ilicis*). The carnation fly attacks glasshouse and outdoor carnations, pinks and sweet william in the early autumn whose leaves are often so heavily mined that they shrivel up. If the shoots are tunnelled into they frequently die. Once the maggots have matured they leave the plants and overwinter in the soil, but some younger ones remain to feed until the end of winter. Both pupate in the spring and the adults which resemble house flies emerge at the end of May to the beginning of June. Two generations of the 6mm long sturdy maggots of the chrysanthemum blotch miner cause linear mines along the leaf veins of *Chrysanthemum maximum* and autumn-flowering chrysanthemums which later become blotches. The adults that emerge in the spring lay eggs which result in the summer larval attack; these will produce the next generation of adults in July. The peg-shaped maggots of chrysanthemum leaf miner attack a range of cultivated and weed hosts outdoors but with little lasting harm. Damage is particularly severe on composites including various chrysanthemums, cineraria, calendula, gazania and helianthus. There can be several generations in glasshouses as the lifecycle is often less than four weeks in duration, but outdoors there are only two. The mines which are most obvious on the upper leaf surface are very sinuous and appear white to brown and can both disfigure and ultimately shrivel so many leaves that young plants die and the quality of the flowers deteriorates in older ones. The marks left by the females after they lay their eggs also blemish the plants. The application of a suitable insecticide should start as soon as these marks appear. Chrysanthemum leaf miner flies often invade glasshouses as the weather outside begins to cool down. The holly leaf miner is common in gardens and in the wild. Infected leaves are disfigured by linear mines that develop into blisters where the yellowish-white maggot pupates overwinter and the small black adult flies emerge at the end of May. Although some insecticides are effective, infested leaves can be picked off and burnt like the leaves affected by the other leaf miners, providing the attack is not too severe.

Left: **Chrysanthemum leaf miner.**
Above: **Leaf spot lesions on *Helleborus niger* leaves.**

Leaf spot

Leaf spot can rapidly spread to all new growth on most *Helleborus* species. The irregular, black leaf spots develop often virtually undetectably at first on a few leaf trusses at early flowering, before enlarging to merge together to form unsightly concentric scorch-like patches. These eventually bear numerous perithecia on both sides within the dead areas on the leaves. Eventually the flower stems and petals are also often covered in these blotches and may shrivel up. However, the leaves are usually more severely affected, and frequently become gradually yellowed and eventually die if both sides of the leaves become covered with blotches. Overall growth can be seriously suppressed and as a result the flower buds may fail to develop properly.

When the leaves start to die the infected plants become increasingly stunted and malformed especially if the tips of young shoots are infected. Following severe stem infection the plant may collapse and even die. Less serious levels of disease may weaken the growth of the plant sufficiently for it to be substandard in subsequent years.

Hellebore leaf spot, *Coniothyrium hellebori*, is transmitted mainly by ascospores released from the perithecia embedded within the surface of the dead leaves during damp weather. Ascospores that discharge in spring from active perithecia renew the infection each year as the hellebores emerge after winter. Ascospores can also be released in the autumn, if it is wet. No actual wound damage is required before infection as the ascospores can readily infect intact leaves, so within a few days inoculum can quickly spread between all the plants in a bed.

Spray with an approved fungicide beginning in the autumn and continuing until all new leaves and flower buds have emerged in spring if heavy attacks have been experienced in the previous season. Try to minimise the build-up of moisture by covering particularly susceptible plants during winter and spring. Cultural methods of control include planting hellebores in well-drained open ground, well spaced and without too much shade in order to encourage open growth. Cutting off and burning very heavily infected leaves can help, but this should not act as a substitute for prevention after severe attacks. Although no *Helleborus* sp. is completely immune, there is some evidence of some varietal resistance in some of the tougher-leaved species such as *H. corsicus* which may escape the worst symptoms, but even they can be badly disfigured during wet springs and autumns.

Mammals

Although some mammals can cause substantial damage in the garden most do not eat foliage but concentrate on debarking trees, damaging plant stems, flowers and buds, and uprooting plants. Deer of various kinds, rabbits and hares graze on grass and other foliage. In the winter they gnaw the bark off trees and shrubs, often killing them. Deer tend to jump fences that exclude rabbits and hares. They have been known to be deterred if human hair is scattered over foliage or the dung of predators, such as lions is applied, if this can be obtained from a local zoo. Landowners should be reminded that they have a legal responsibility to control the rabbits on their land. Voles can cause damage during those years when their populations become numerous, but they can be discouraged by removing the grass or other plants that grow around their nests.

Mealybugs

Mealybugs are similar in some ways to scale insects, to which they are related. However, they have no actual scale covering – the females and immature stages are covered by a powdery white wax coating. Also, mealybugs are mobile and attack a wide variety of plants whose sap they suck. Plants that are affected are generally stunted and lose leaves, which turn yellow before dropping.

Mealybugs are flattened, elliptical wingless insects, a few millimetres long, that are soft to the touch. Most are pale yellow or pink with a fringe-like covering of whitish waxy structures around their bodies, particularly the tail end. The eggs are also covered in wax.

Since mealybugs produce copious amounts of honeydew, it is this, or rather the sooty moulds that grow on it that attract attention, as the mealybugs themselves are usually are found hiding within curled leaves, leaf sheaths or buds. Often the mealybugs cannot be sprayed easily even when they can be seen, for example, when they cluster around the spines of cacti. Most mealybugs are confined to glasshouse plants but one, the currant mealybug (*Pseudococcus fragilis*), frequently attacks ceanothus, flowering currants, forsythia, laburnum, robinia and spiraea, as well as many glasshouse plants. This mealybug is most frequent in the southwest of England.

Some mealybugs are found underground feeding on the roots of plants. These mealybugs are often found with ants which protect them from being attacked by predators in return for a steady supply of honeydew. Although several species attack outdoor plants, they are rarely serious enough to warrant control, unlike some species that attack glasshouse plants. The same insecticides used against scale insects are usually effective against mealybugs as well. However, it is usually rather more difficult to spray mealybugs so it may be worthwhile dabbing on the diluted spray insecticide with a fine paintbrush.

There is only one really important sucker in Britain – the box sucker (*Psylla buxi*). In summer, adult box suckers look like green-winged aphids with brown markings, but they darken in autumn.

***Above*: Young tree damaged by deer.**

Although they are somewhat similar in size (3mm) and colour to many common aphids, the adults are rather more stockily built, yet at the same time they are more vigorous. The adults can move quickly by jumping as well as flying. The eggs, originally laid in August, hatch in spring and the prostrate, yellowish to dark green nymphs invade the opening buds of box. Unless they are interfered with, the crab-like nymphs stay hidden inside the leaves of the bud which have curled over them where they feed on the shoots before developing into adults at the end of April. Infested buds resemble miniature Brussels sprouts from which the nymphs excrete tubes of honeydew inside waxy tubes. These break easily, releasing the sticky honeydew which can become colonised by dark black, sooty mould fungi. As a result the box bushes become rather shabby.

A systemic insecticide is necessary to penetrate through the curled leaves and kill the nymphs, but some conventional insecticides can be effective if they are applied before the adults start to lay eggs.

Mites

Although there are exceptions, most adult mites, which are tiny relatives of the spiders and other arachnids rather than insects, walk on eight legs, but their larvae often have only three pairs. Other differences are that they are unsegmented and have no distinct head – this is merged with the rest of the globular body – nor wings and antennae. The four main groups of mites that live on plants (red spider mites, tarsonemid mites, acarid mites and gall mites) have evolved piercing mouth parts that suck sap.

In general adult female red spider mites are oval, between 0.3-0.7mm long and, with the smaller, more elongated males, are found feeding on the undersides of leaves along with the larvae and the two nymph stages. However, in some species such as the bryobia mite (*Bryobia* spp.) males are unknown because reproduction takes place by parthenogenesis, with five to eight generations occurring a year. Bryobia mites are relatively large, flat, red spider mites with rather long front legs, that actually vary in colour between light green and reddish-brown. Bryobia mites often live on the upper surfaces of leaves. They can be found during sunny weather from April until the time they overwinter in crevices, but the mites hide when they are not feeding. Although rarely a serious pest, bryobia mites attack a wide range of ornamental plants, including grasses, and some flowering trees. The initial symptom is a fine speckling which progressively spreads until the leaves turn yellow, bronze or silver. Often enormous numbers of bryobia mites are found in new houses in the spring, where they have hatched from eggs laid by overwintered adults that were attracted to the

***Left*: Bryobia mite on ivy leaf.**
***Above*: Severe long tailed mealybug infestation on glasshouse palm.**

plaster and mortar while it was still moist. The best way of controlling these invasions is to move infested plants away from the house and replace the soil with gravel.

The fruit tree red spider mite (*Panonychus ulmi*) is found on many flowering trees such as cherry, almond and crab-apple as well as flowering currants, especially if these have been sprayed with insecticides that do not harm the mites but kill off their predators. The dark red, adult, female fruit tree red spider mites are about 0.4mm long and larger than the males, some of which are red but many are light green. Four or five generations of these feed on the undersides of leaves each year, together with the mites in the larval stages. The first generation hatch from the large clusters of bright red eggs that were laid in the bark in September where they will overwinter, but the later generations result from the orange-red summer eggs that are laid on the undersides of the leaves. Fruit tree red spider mites are traditionally controlled on fruit trees by a winter wash of petroleum oil. Glasshouse red spider mites (*Tetrancychus urticae* and *T. cinnabarinus*) are very serious pests of a great variety of glasshouse plants especially under hot, dry conditions, where they feed on the undersides of leaves with the immature stages. The affected foliage becomes more and more speckled until the yellowed leaves turn rusty and heavily infested plants will often die. Red spider mites produce fine webs which frequently link the leaves on which they feed. Even though they are superficially similar in many ways, with the females larger and more egg-shaped than the males, the two species of red spider mites that are found in glasshouses – *Tetrancychus urticae* and *T. cinnabarinus* – are somewhat different in behaviour. Whereas *Tetrancychus urticae* is found outdoors on a wide variety of ornamental and weedy hosts, *T. cinnabarinus* is more restricted to glasshouses. The young of both are yellowish, but there are some differences in adult coloration. Adults of *T. urticae* are usually light green but the females become brick red when starved or entering hibernation. The females of *T. cinnabarinusare* normally dark red and the males yellowish. In the glasshouse both types produce several generations each season, yet except on year-round chrysanthemums, *T. urticae* usually stops breeding at the end of summer before hibernating in crevices until spring, while *T. cinnabarinus* continues to reproduce all year. The females lay up to 100 eggs on the lower surface of the leaves protected by the webs until they hatch 32 days later. Those of the former are translucent whitish and those of the latter are pale pink. Many strains of both species are now resistant to the pesticides that were formerly used against them, so one of the most effective methods of controlling both species of mites is by introducing the predatory mite, *Phytoseiulus persimilis*, which feeds exclusively on red spider mites and can eat up to six adults or

20 of their young each day. Although similarly sized, *P. persimilis* is pear-shaped with longer legs and is more active than its prey. The predator should be used as soon as it is available and at the first signs of attack. It is a good idea to group the infested plants together. Control can be improved by keeping the glasshouse at between 16 and 30°C and pretreating the plants by spraying them with water, as red spider mites prefer warm, dry conditions. The predator does not usually escape through glasshouse vents in any substantial quantity, but it cannot survive without a population of red spider mites and so may have to be reintroduced regularly, especially when accidentally eradicated by inappropriate pesticides. For this reason, fatty acid insecticides are often recommended as a pretreatment, although a day still has to be allowed before the predator is introduced. Often two introductions are necessary under glass, but reintroduction will usually have to be repeated more often if the predator is used outdoors, as it cannot survive if the weather is too cold. It is wise to try to prevent *T. urticaefrom* hibernating in crevices by keeping the lighting on while applying control measures including removing the old crop with its mite infestation intact while the crop is still alive. It is always worth trying to remove weed hosts from the area.

Tarsonemid mites, such as the broad mite (*Hemitarsonemus latus*), the bulb scale mite (*Steneotarsonemus laticeps*), the cyclamen mite (*Tarsonemus pallidus*) and the fern mite (*Hemitarsonemus tepidariorum*) are so small (0.25mm) that their presence is usually not detected until damage is seen. The male tarsonemid mites are even smaller than the females, but nonetheless are often seen carrying around a quiescent six-legged female larva with which they intend to mate as soon as it matures. This behaviour also encourages spread, as the males generally transfer their intended mate to new leaves. However, in other species there are far more females than males and reproduction is usually parthenogenetic. The broad mite is a tiny, whitish translucent mite that attacks a wide range of glasshouse plants to cause stunting. It cannot easily be distinguished from the cyclamen mite, but is more easily controlled by pesticides. It is found on the underside of leaves, particularly those that are young and tender, which it causes to become shiny and distorted. The margins of the leaves curl down and the undersides may become bronzed, especially in begonia. In other species such as chrysanthemum and gerbera the flowers are deformed, or drop off in the case of impatiens. Broad mites continue to breed throughout the year; the life cycle is often completed in four days in the heat of summer. Several pesticides are effective and are often also applied to the soil to control mites that fall from the foliage. The cyclamen or strawberry mite produces similar symptoms to the broad mite on a similar range of plants but flowers often display streaks and blotches. Cyclamen mites can live outdoors on the strawberry plant and Michaelmas daisy. Under glass the cyclamen mite has a somewhat longer live cycle than the broad mite, but outdoors they overwinter in the dead flower heads of Michaelmas daisies where they continue to breed if it is mild. Apart from the few pesticides that are available, it is sensible to remove and burn all dead material from Michaelmas daisies. Fern mites can seriously distort the fronds of *Asplenium bulbiferum*, and can disfigure *Pteris* and *Polystichum* spp. by feeding on the young fronds and their bases. Eggs are laid throughout the year, but during the winter often only females are present. Fern mites are generally controlled by pesticide sprays.

The bulb mite (*Rhizoglyphus echinopus*) is the only important acarid mite. This pest is covered in the section on bulb pests and diseases along with the tarsonemid mite, the bulb scale mite (*Steneotarsonemus laticeps*). Gall mites are tiny and semi-transparent, with only two pairs of legs next to the head. Little is known of their life cycles, but it is likely that they remain hidden inside the bud scales over winter until they become active in the spring as plant buds open. In general, gall mites only feed on the surface of leaves and so are not very harmful, but there are two destructive glasshouse species, the chrysanthemum leaf mite (*Epitremerus alinae*) and the *Paraphytotus* sp., which can be damaging to cuttings. Although these species cause typical mite symptoms on leaves, including distortion and leaf fall, most of the other species only cause galls – abnormal growths which spoil the appearance of the plants. Such galls can be divided into various types including the felt galls seen on birch and other trees; pocket galls seen on pear, lime and maple; leaf roll galls seen on hawthorn; and bud galls seen on blackcurrant, hazel, yew, birch and broom. However, control measures are rarely warranted, except against bud galls on blackcurrant where insecticides are often used, as overall growth is usually unaffected.

Powdery mildews

The key feature of most powdery mildews is that severe infection can eventually weaken the plant, especially where variety and local environmental conditions are inappropriate. The severity of infection is often patchy due to differences in microclimate. Apart from intracellular haustoria which penetrate the epidermis, the mycelium is restricted to the outside of the leaves. This whitish covering of hyphae bears dense groups of concealed brown, fruiting bodies (the cleistothecia) as well as the much more abundant and noticeable initial powdery chains of conidial pustules which eventually merge as felt-like patches over the leaves and, where present, the twigs and fruit. However, the importance of the cleistothecia in ensuring the survival between seasons is considered to be less significant on many woody plants than dormant mycelium overwintering in the buds.

Apple powdery mildew

Podosphaera leucotricha infected leaves are covered with a white powdery mildew. Often infected leaves drop off so twigs bear leaves only at their tips. Sometimes the surface of apple fruit can also be spoilt by mildew. Edible and ornamental apple (*Malus* species), peach, quince (*Cydonia vulgaris*) and *Photinia* spp. are all susceptible. Infection from previous years survives over winter in buds which can be recognised as they are narrower and less plump than healthy buds, with narrower outer scales peeling back. Infected buds open rather later in the spring than those which are healthy to reveal shoots conspicuously covered in powdery mildew. These shoots remain stunted during the growing season, often losing the lower leaves. Any flowers present stay greenish but ultimately fall off without setting fruit, thus reducing the yield of fruit. Prune out and burn twigs harbouring dormant mycelium overwintering in the buds of infected trees (sprays of a suitable winter fungicide wash have previously been effective but are no longer available). Fertilize, water and prune healthy trees early to stimulate vigorous growth by providing sufficient potassium and regulating the nitrogen supply. Several fungicides are available for garden use.

***Top*: Powdery mildew on apple shoot.**
***Above*: Powdery mildew on young rose leaves.**

Rose powdery mildew

Sphaerotheca pannosa is extremely common and widespread worldwide on Rosa spp. and some other members of the Rosaceae. Mildew develops most extensively on leaves and young shoots, but these become resistant on maturity. Petals, sepals and receptacles of flower buds may also be invaded. Infected buds frequently remain closed. Infected leaves sometimes become yellow or purple. Roses in glasshouses or dry sunny positions, such as against walls, are particularly susceptible to the disease and may be killed. Elsewhere the consequences of the main initial spring infection by wind-blown conidia are not normally seen before May. Two biologically distinct strains of *Sphaerotheca pannosa* (Wallr.) have been reported – one on rose, another on peach. In mild seasons, hyphae overwinter as whitish felt-like mats on leaves, calyx, fruits and shoots, especially around the thorns, or in dormant buds. Few cultivars listed as resistant are consistently reliable everywhere. Some varieties formerly considered very resistant are now often susceptible in some gardens. Cultural methods, such as pruning out disease, limiting the use of nitrogenous fertilisers and using mulches to provide extra moisture on dry sites, may need to be followed by a regime of approved fungicide sprays, often reapplied fortnightly.

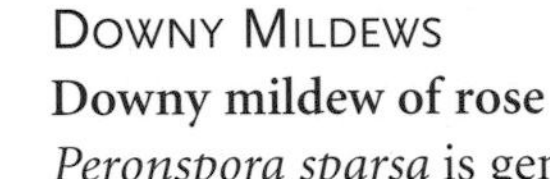

Downy Mildews

Downy mildew of rose

Peronspora sparsa is generally found on glasshouse roses rather than in gardens. Like other downy mildews it tends to be a problem under damp conditions. Unlike the unrelated powdery mildews, downy mildews are inhibited by dry growing conditions and prefer the undersides of leaves. The sporangia that are borne on the branched sporangiophores are released under damp conditions which favour both the germination of the sporangia and that of the motile zoospores which burst from the sporangia and swim through films of water to cause new infections. Downy mildews also produce a resistant spore, the oospore, which remains in the rotten mass of the plants that they have attacked. Both zoospores and the mycelium in the rotten tissue ensure that the downy mildews survive overwintering. In the spring the oospores and the mycelium form fresh sporangia and the cycle continues. Rose downy mildew can be detected as it causes dingy russet-coloured patches to form on the upper leaf surface. If the leaf is turned over the presence of off-white sporangia is revealed bristling on the underside of these lesions. Heavy infection may cause the leaves to drop off. The most effective way of treating rose downy mildew is to make sure that the glasshouse is properly ventilated especially during the night, but the rose will also respond to appropriate fungicide sprays.

Downy mildew of wallflower

This downy mildew, *Peronspora parasitica*, is also found on alyssums, many vegetable crucifers such as cabbages, turnips, radishes and many others as well as several weeds including shepherd's purse. The symptoms are first seen on seedlings, where the off-white sporangia are found on the underside of leaves that have speckled yellowish lesions on the upper surface. The fungus spreads to other seedlings through the soil and infects them systemically through their roots. Sporangia from the seedlings spread in the air by wind to neighbouring plants. Some seedlings will be killed by these lesions, while others, although distorted, will survive to become mature plants. Mature wallflower plants suffer similar lesions but by and large, they are much less seriously affected. It helps to increase the ventilation but an appropriate fungicide may be required.

Rhododendron gall (*Exobasidium species*)

Most frequent and prevalent on rhododendron bushes in many north temperate regions of Europe, Asia and America with suitably mild moist climates or in artificial environments that provide warm, moist conditions. Glasshouse azaleas are often regularly attacked. Such aerial infection may be localised and cause small irregular swellings on leaves, buds or flowers which later either become enclosed as leaf spots or hypertrophied. Leaves, flower buds and shoot tips may become malformed. Leaf concavities at first thicken, redden above a yellowish margin, then turn pinkish white and finally chalky white as a covering of basidiospores develops on bud or leaf galls and hypertrophied leaf spots. Dissemination is by airborne basidiospores and conidia, probably by rain splash, on hands, tools, clothing and possibly also by insects. Usually only young leaves become infected by direct penetration of the cuticle. The mycelium spreads throughout the cortex and into the pith of infected stems and may become systemic. Isolates of strains from different host species show some physiological differences and may not be capable of cross infection. Many aspects of the pathology of this pathogen have not yet been fully investigated. Although a number of fungicides have been tried, the disease is most effectively treated by removing individual galls. This should preferably be done before they turn white and are able to infect healthy flowers and foliage. Cuttings should not be taken from plants that show symptoms.

Rusts

(Order *Uredinales*)

These basidiomycete fungi debilitate their hosts by hyphae that develop within the host tissues and their haustoria (special feeding structures) that drain nutrients from the plant without killing it. After a time these produce areas of fungal growth hidden inside leaves and stems, which then burst though as rust-like pustules. Many different rusts, some very common in gardens, infect a wide range of vegetables, fruit, trees and ornamentals. Most rusts are *Puccinia* species, some of which can have as many as five different spores (macrocyclic), some may switch between different hosts (heteroecious), many are much simpler and only have some of the spore stages (microcyclic) and some are only found on one host (autoecious). Pycniospores may be produced in insignificant flask-shaped structures typically on the upper surfaces of leaves, often with a sweet secretion that attracts insects which move to the receptive hyphae of compatible structures. After fusing together, a conspicuous group of cluster cup (the aecia) structures may often be found partly implanted beneath the lower

***Left*: Downy mildew (blue mould) on *Nicotiana* leaves.**

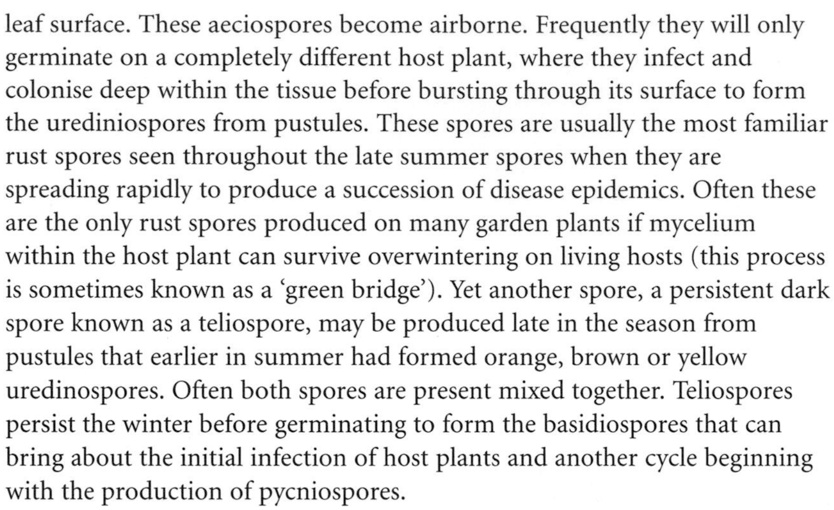

leaf surface. These aeciospores become airborne. Frequently they will only germinate on a completely different host plant, where they infect and colonise deep within the tissue before bursting through its surface to form the urediniospores from pustules. These spores are usually the most familiar rust spores seen throughout the late summer spores when they are spreading rapidly to produce a succession of disease epidemics. Often these are the only rust spores produced on many garden plants if mycelium within the host plant can survive overwintering on living hosts (this process is sometimes known as a 'green bridge'). Yet another spore, a persistent dark spore known as a teliospore, may be produced late in the season from pustules that earlier in summer had formed orange, brown or yellow uredinospores. Often both spores are present mixed together. Teliospores persist the winter before germinating to form the basidiospores that can bring about the initial infection of host plants and another cycle beginning with the production of pycniospores.

If the rust is an heteroecious species that interchanges successively between different host plants, the pycniospores and aeciospores are usually produced on one host species, called the primary host, and any other spores on another. The two host plants are often quite unrelated so some rusts may be controlled if one of the less valuable alternate hosts is destroyed or removed.

Avoid nitrogenous fertilisers which aggravate the effects of many rusts. Some fungicides control rusts if applied as protectants or systemic fungicides. A plant variety that is resistant to an individual strain or race of rusts is rarely resistant to every rust strain present in the area. Although plant pathologists refer to the plant on which the aecia are found as the primary host, rusts are often named after the most important host on which the most conspicuous pustules are present. These can be cluster cups, uredinospores or teliospores.

Antirrhinum rust

Puccinia antirrhini spread from western North America in the 1930s and is common wherever antirrhinums are grown, especially in areas prone to drought. This rust is microcylic as it only produces brown pustules of uredinospores and sometimes pustules of teliospores are also formed. The pustules of urediniospores are surrounded by a yellow halo. Usually infected plants lose most of their leaves and the flowers do not form or are malformed. Several races of the rust ensure that all varieties of antirrhinum are susceptible. Some fungicides are effective, but are rarely totally effective because the disease survives as mycelium inside the host plants. These plants are able to overwinter the rust and so should be uprooted and burnt. Seed may also be infected, although infection is lost if the seed is stored longer than a year.

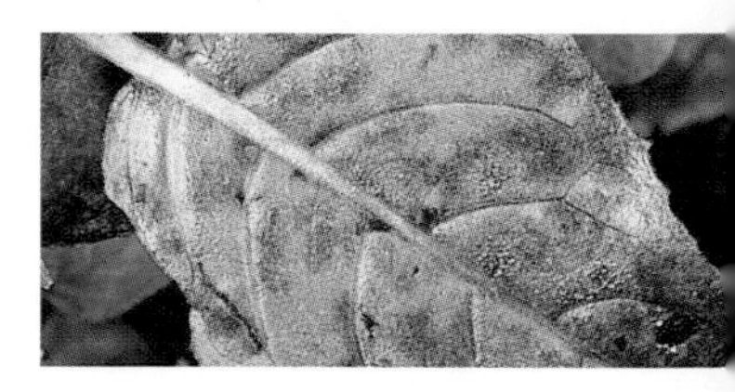

Fuchsia rust

Although *Pucciniastrum epilobii* is not very common it can be encountered if fuchsias are grown near willowherbs. If a fuchsia leaf is infected, the upper surface is marked by a pale yellow patch above an area of light orange urediniospores on the leaf underside. Heavily infected leaves often become withered and die.

Mahonia rust

This rust, *Cumminsiella mirabilissima*, originated in North America and has spread onto plants of *Mahonia aquifolium* and *M. bealii* throughout Europe. It can be detected on the upper surface of the leaves as sharply defined dark reddish spots which lie above the powdery pustules beneath. These pustules generally consist of urediniospores and later teliospores, but in early summer aecidiospores may form underneath the leaves in cluster cups. All infected foliage should be pruned out. Repeat fungicide sprays fortnightly.

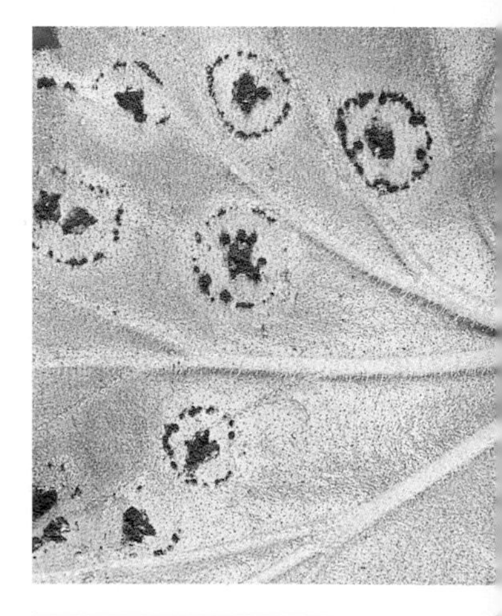

Pelargonium rust

Puccinia pelargonii-zonalis is native to South Africa, this rust is spreading to wherever susceptible varieties of *Pelargonium zonale* and its hybrids are grown in glasshouses and other sheltered artificial environments or outdoors as bedding plants. Physiological specialisation is unknown. The rust has spread rapidly in Britain since 1965, reaching as far as Glasgow by 1970, and is now common under warm conditions throughout Europe, Australasia and Africa. Neither ivy-leaved nor scented pelargoniums are affected. Infected plants may be reduced in size and vigour, only able to produce a few flowers and new shoots. Like many other rusts, the most obvious signs of infection are the yellowish spots that form on the upper surface of the leaves, marking the place beneath which rusty cinnamon-brown pustules, the powdery uredosori (uredia), have erupted. Cuttings should not be taken from plants that show symptoms. Heat treatment of cuttings has been investigated but can result in damage if treatment is not fully controlled. Commercial growers use fungicides, some of which are not approved for amateurs, but several general garden fungides are also effective. Other control measures suggested are hard winter pruning, removing diseased leaves as early as possible before they produce uredinosori and are

Top left: Rust on Sweet William leaves.

Left: Antirrhinum rust symptoms.

Above: Fuchsia, Mahonia and Pelargonium rust.

able to infect healthy foliage, and reducing humidity, by ventilation in the greenhouse or moving unaffected plants to a drier position indoors.

Periwinkle rust

Puccinia vincae is not very common but is easily identified as infected plants do not usually produce flowers and instead of scrambling over the soil, the shoots with distorted leaves grow upright. Although the five different rust spores are produced, the brown aecidiospores form the most conspicuous pustules. Since the rust forms a perennial mycelium that invades the rootstock it is best to pull up and burn any diseased plants.

Mallow and hollyhock rust

Puccinia malvacearum was first recorded in Chile in 1852, then Australia, before it was reported from Spain in 1869, after which it spread rapidly throughout Europe. Outbreaks of hollyhock rust are very common in gardens. In addition to hollyhocks, wild and ornamental mallows are also infected. The initial signs of hollyhock rust are the pustules that erupt on the stem and through the leaf cuticle. These russet-brown pustules are generally on the lower surface of the leaves but associated with small yellowish patches on the other side of the leaf. As a result of these lesions the plants wilt and become almost completely defoliated, yet rarely die. *Puccinia malvacearum* is a microcyclic rust which is autoecious. This means that is it is able to complete its life cycle with only two of the five possible rust spore states on one host, generally species of hollyhock (*Althaea*) or mallow (*Malva*). However, instead of the urediniospores that are so often the only spore stage present, in this case the spores known as teliospores are produced in persistent dark pustules in summer. Later in the season these appear whitish as they germinate to bear the basidiospores that can bring about the initial infection of new host plants. Teliospores can survive on plant debris, but mycelium generally also survives over winter in an intimate association within the hollyhock or wild mallows nearby. However, the rust mycelium weakens its hosts somewhat by developing extensive intracellular hyphae with haustoria that drain nutrients from within the cells of the plant tissues without killing them. It is better to grow new plants from seed well away from the old, as the infection is systemic in the plant once established. Avoid nitrogenous fertilisers which intensify the effects of hollyhock rust on its host plants. Some fungicides control hollyhock rust if applied as protectant or systemic fungicides with a wetter to overcome problems of wetting the hairy leaves. Varietal resistance is unreliable and only partially effective as a plant variety that is resistant to an individual strain or race of hollyhock rust is rarely resistant to every rust strain present in the area. Eliminate alternate wild mallow hosts from around the garden.

Rhododendon rust

This rust, *Chrysomyxa ledi* var. *rhododendri* has two hosts. The primary host on which the aeciospores are formed is a spruce, usually Norway but occasionally Sitka. On the underside of the leaves and on the young green shoots of rhododendron the urediniospores are formed as orange patches. These could be confused with the activities of the rhododendron bug but are easily confirmed if a microscope is used. The symptoms on the spruce are yellow bands which later bear white aecia. The aeciospores can even infect the relatively resistant common purple *Rhododendron ponticum* but only if it is found nearby. On other more susceptible rhododendron species it is usual for the urediniospores to be produced perennially. Although some fungicides are effective, it is wise to remove and burn infected leaves as soon as they are seen.

Rose rust

If your garden is rather damp you may see bright yellow patchy areas on the upper surface of a rose leaf. Turn the leaf over and you will probably find the powdery orange or black pustules of rose rust beneath them. The orange pustules become black in the autumn. If you notice rose rust, choose an approved fungicide that controls it as well as the more common powdery mildew and black spot.

Although the identity of the rust fungus responsible is sometimes still disputed, *Phragmidium mucronatum* is common and widespread throughout Britain and northern Europe. The initial symptoms which arise in spring are the bright orange pustules on leafstalks, branches, the undersides of leaves (especially on the veins) and any remaining rose hips. The aeciospores that develop from these pustules germinate and infect leaves producing yellow-orange pustules that bear the more familiar urediniospores on the undersides of the leaves during the summer. Towards the end of summer these orange pustules gradually turn into black pustules. The teliospores that are produced in the black pustules are able to survive the winter on fallen leaves. In spring they germinate to produce basidiospores which infect the new leaves after being spread by wind or rain splash. Although most species and cultivars of roses are susceptible, a few show some resistance. In the autumn it helps to collect up and burn fallen

***Left*: Hollyhock rust.**

rose leaves. A number of garden fungicides are all usually effective if sprayed fortnightly during the summer. It is sensible to use a fungicide that is also active against black spot and powdery mildew if these also occur.

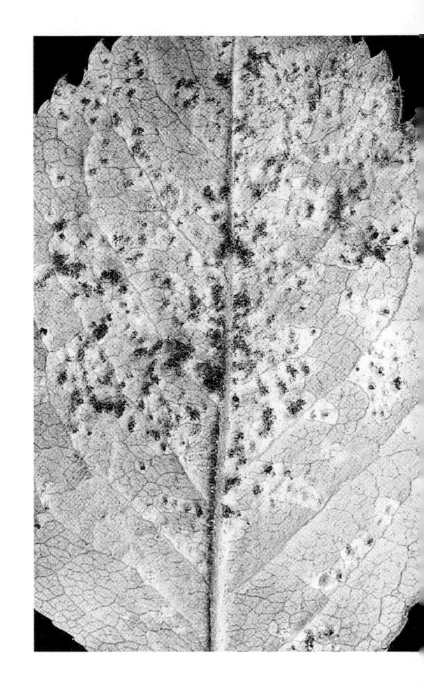

Sawflies

Sawflies are related to the bees and wasps but do not have a narrow waist. They are called sawflies because the structure adapted to cut into the host plants while laying eggs is serrated. The larvae that hatch resemble caterpillars, but in addition to the six pairs of legs on the thorax they have six or eight pairs of prolegs instead of five. Nearly all have a brown head. Most sawfly larvae that attack herbaceous plants feed on leaves. The larvae of several species are important pests and include the antler sawflies (*Cladius pectinocornis*) which are flat and green, usually found feeding around the edge of rose leaves. The adults emerge from cocoons on the undersides of the leaves or leaf litter in August and produce a second generation that overwinters as a cocoon to emerge in May. The banded rose sawfly (*Allantus cinctus*) is also found on roses but has spots on its back, is greyer and paler green on its underside. Unlike the antler sawfly, the banded rose sawfly initially feeds on one surface of the leaf and only later starts to devour irregular holes through it. There are also two generations per year, though the larvae burrow into the plant stems to pupate. The large rose sawfly (*Arge ochropus*) cause two damaging symptoms on roses. The young shoots and flower stalks darken and twist where the eggs are laid in rows of slits. Two generations of the yellow-backed greenish caterpillars with black spots at first devour one leaf surface and then bite through it to the other side or eat away the leaf margin.

The leaf rolling sawfly (*Blennocampa pusilla*) attacks several varieties of bush and climbing roses but rarely affects standard roses. The leaves which roll up towards the midrib are those which were gouged prior to egg laying, and many but not all form a protective shelter for the pale green caterpillars; in others the eggs may not have been laid. Even where there is no larva, if too many leaves are affected the plant may soon become less vigorous.

The translucent yellowish larvae of the rose slug sawfly (*Endelomyia aethiops*) resemble slugs and are known as slugworms. These larvae pare off one surface of a leaf which then turns mottled brown on the other side. Once they have become fully mature the slugworms overwinter in a cocoon, though some may remain there for a year and a half until the subsequent season. Rose bushes can be weakened by heavy infestations. There are two geum sawflies (*Metallus gei* and *Monophadnoides geniculata*). The larvae of the former are leaf miners which are found communally in blotch-type

Left: Pear and cherry sawfly larvae on damaged cherry leaf.
Above: Rose rust on underside of leaf.

mines which connect up to affect almost the entire geum leaf surface, but the dark green caterpillars of the latter voraciously eat away holes in the leaves of geum and filipendula. While the former have two generations, the latter have only one. Two generations of the lead grey larvae of the iris sawfly (*Rhadinoceraea micans*) eat the leaf margins and sometimes the flower buds of various wild and cultivated iris. The slug-like larvae of the pear slug sawfly (*Caliroa cerasi*) attack several roseaceous trees and shrubs. At first they are pale, darkening with age but eventually become yellow before pupating in the soil. These slimy larvae, the slugworms, eat away the upper surface of the leaf which then shrivels and dies. A second generation causes damage in late summer before overwintering as cocoons, but some pupae produce the start of a third generation. Two and sometimes three generations of the shiny greenish larvae of the spiraea sawfly (*Nematus spiraeae*) feed in groups between the veins on the lower surface of the leaves of goat's beard (*Aruncus sylvester*) until only a skeleton remains.

It is fairly easy to control most sawfly infestations with insecticides. However, those of the leaf-rolling sawfly and the geum leaf miner are too well protected and so it may be more effective to remove the rolled or mined leaves by hand and burnt if they are not too numerous.

Scale insects

The majority of adult scale insects are found on the bark of trees or shrubs. Nonetheless, a number of the scale insects that are more commonly found under glass or in sheltered gardens do colonise leaves. The cushion scale (*Chloropulvinaria floccifera*), cymbidium scale (*Lepidosaphes machili*) and the orchid scales (*Diaspis boisduvalii*) are found on orchid leaves. The cushion scale produces a characteristic, long, white egg sac and the cymbidium scale resembles the mussel scale but is only found on Cymbidium orchids. The female orchid scales, which are round flat and translucent, occur on several genera of orchids and palms. The female fern scales (*Pinnaspis aspidistri*) are mussel-shaped and infest aspidistra, fern and palms. fluted scale insects (*Icerya purchasi*) attack citrus, acacia, cytisus and mimosa. Hemispherical scales (*Saissetia coffei*) are a problem on begonia, carnation, clerodendrum, croton, ferns, figs, oleander, ornamental asparagus and stephanotis. Soft scale (*Coccus hesperidium*) is common both under glass and outdoors in sheltered gardens as it infests a very wide host range, usually on the undersides of their leaves in layers along the midribs. Female oleander scales (*Aspidiotus hederae*) are found on many herbaceous plants, palms and woody shrubs, often including *Aucuba japonica* outdoors, but few of the species that are hardy enough to survive outdoors regularly infest leaves except the juniper scales (*Carulaspis juniperi* and *C. minima*) which deform and blemish cypress, juniper and thuja.

Many species reproduce by parthenogenesis. However, if these appear, the tiny adult winged males only survive long enough to mate with the scaled females. In some species the eggs are retained on the plant surface held under the female's scale even though they hatch after she is dead. The eggs in other species are covered by a waxy wool-like material. Young nymphs known as crawlers leave the protection of the dead scale or wool after they hatch and try to discover acceptable sites where they can develop into either females with permanent scales or males which relinquish their scales before mating.

Often the scale insects that are found on leaves are easily missed as the adult winged males – if they occur – and the young are tiny, and the larger females and young males are often completely concealed by the scales that protect their wingless bodies, remaining constantly fixed to their host plants. When large numbers of scale insects are feeding on sap from leaves they usually induce substantial injury by both stunting and yellowing the foliage. Typically the leaves of the affected plants are also thoroughly spoilt by a dirty covering of sooty mould fungi that grow on the honeydew that the scale insects excrete.

Some insecticides are recommended as sprays but chemical sprays are only partially effective apart from the tar oil sprays to dormant woody plants. This is because the female scale insects and their eggs tend to survive under the sturdy waterproof scales. Inspect all new plants for scales, and as a precaution quarantine them in an isolated place for a few weeks. If they are found to be infested the scales should either be detached by scratching them off or treated by daubing them with a cotton-wool bud soaked in an appropriate dilute insecticide.

Slugs

Slugs are often the most damaging of all the garden pests to a wide variety of garden plants, particularly in the seedling stage. As many as 200 slugs can live in a square metre of garden and as well as attacking the foliage, most of these slugs also devour the roots of many plants. Slugs like damp conditions and so are much more common in wet years and in areas of the garden that are humid, particularly from March to October. They are favoured by loam and clay soils which retain water as well as oils with a high humus content, but are less common on highly organic peaty soils. Slugs tend to hide away under rank vegetation, wood or stones during the day to emerge at night to feed providing the conditions are sufficiently warm and humid. They make

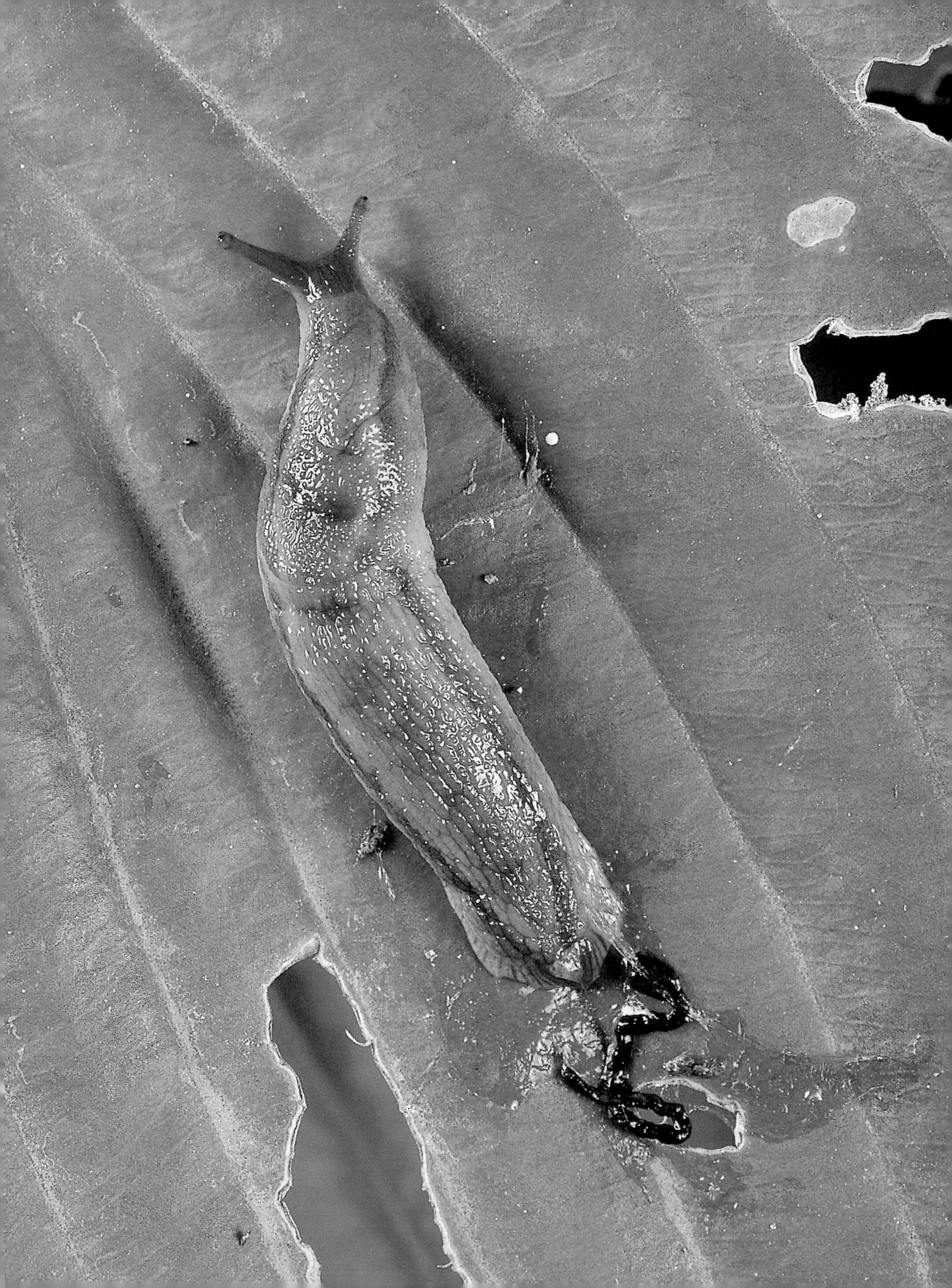

irregular holes in the leaves, flowers and stems with the minute teeth that cover their rasping tongue. This feeding soon results in making the plant very unsightly. Several species of slugs are common in gardens, of which the three important types in Britain are the field slug, *Deroceras reticulatum* and some related species; round-backed slugs such as the garden slug, *Arion hortensis*; and keeled slugs, *Milax* species, which tend to attack plant roots. All of these need mild damp conditions in which to breed, and so they avoid dry areas. Gardens generally provide slugs with both an appropriate microclimate as well as plenty of living plants and dead organic matter on which to feed. During frosty or dry weather slugs bury themselves deep in the soil and hibernate over winter. Most slugs have similar life cycles. Several hundred eggs are laid annually in batches of ten to 50 most commonly during spring or sometimes autumn by an hermaphrodite parent; these are the result of cross fertilisation by another hermaphrodite in the autumn or winter. Most of the translucent eggs are laid during the spring, usually in damp soil or rotting plant debris well away from cold and dry air. The eggs can hatch fairly quickly into miniature replicas of the adults, but often this process is delayed until the weather improves. As a result in some species such as the garden and keeled slugs the complete cycle can take two years, but in the field slug takes a year.

The rather variably coloured, brown to cream, *Deroceras reticulatum* is probably the most destructive of the surface-dwelling slugs that predominantly attacks foliage but can also damage seeds and vegetative planting material. It can reach nearly 4cm long. Although the black slug, *Arion ater*, is very much larger (14cm) and more obvious than the garden slug, *Arion hortensis* which is also black, it is relatively harmless. Slugs are difficult to control and attempts at eradication are rarely if ever successful in heavily infested gardens. Although slug poisons can be sprayed, for many years the most popular method of control has been to scatter pellets of a bran bait poisoned with metaldehyde or methiocarb. However, their effectiveness is reduced if the pellets become wet during rain, so often the bait is kept sheltered underneath a piece of tile, slate or broken flowerpot. Other baits include beer or milk which are poured into traps which drown the slugs. Still other traditional baits include baiting with half orange skins or pieces of other vegetable material or even sacking which are inspected regularly and any slugs present are destroyed. More recently a nematode parasite of slugs, *Phasmarhabditis hermaphrodita* has been discovered. This is being developed into an effective and environmentally safe method of biological control. It has been tested to make sure that it does not affect earthworms and other animals and plants. It does attack beneficial water snails so the area around ponds should be avoided. Although it works best

***Left*: Garden slug on damaged leaf surface.**

during the warmer summer months, as it is killed at 35°C it cannot survive in birds and other warm-blooded animals. The nematodes enter the saddle-like mantle on the back of the slug, where they release bacteria. These bacteria quickly multiply and provide food for the nematodes which multiply causing the mantle to swell as the result of by-products produced during feeding. This injury stops the slugs from feeding and they burrow into the soil to die after becoming completely invaded by the nematodes. The soil becomes contaminated with the larvae of the nematode, thus escalating the effect of biological control. A couple of products formulated in clay mud are available in Britain by mail order for use as drenches in open soil or containers. The contents of the sachet that arrives within ten days should be stirred into water then further diluted and applied via a coarse-rose watering can. The soil that is to be treated should be moistened prior to treatment and kept damp but friable afterwards. Take care to clear away plant debris and mulches that could limit soil penetration by the nematodes. It is best to treat a week before sowing or planting to allow the nematodes time to find and infect the slugs. These products are marked with a sell-by date but are best used immediately although they can be kept refrigerated for a few weeks. An application gives protection for at least six weeks but the residual effect on the population of slugs lasts several months longer. It is essential that the soil is damp but not waterlogged before the drench is applied and during the following six weeks while the treatment is working. The drench is best applied in the evening between March and October as it is most effective between five and 20°C but can even infect after frost. These dates should be modified in cooler or warmer areas. The nematodes do not control mature land snails as they tend to live in conditions that are too dry for the nematodes.

Snails

Land snails like slugs belong to the class *Gasteropoda.* Snails resemble slugs but possess a coiled shell into which they can retract for shelter and protection. Shells can be very cumbersome to snails but they form an adaptable shelter which, although it is dead and inert, can be enlarged continuously by adding more growth to the spiral without altering its overall shape. Growth can be irregular so that growth lines form on the shell where growth has been slower or even stopped altogether for a period. A newly hatched snail already has a small shell formed inside the egg. Slugs can be considered to be snails that have lost their shells. Both are molluscs which have soft, thick-skinned bodies that taper from the head with tentacles towards the tail and both move slowly by waves of contraction across a path of lubricating mucus secreted by a foot composed of a thick, muscular, slimy pad surrounded by a foot-fringe. In the morning their presence the previous night is often revealed by a trail of mucus showing where they have moved across plants, and other surfaces such as concrete or glass. As this method of moving entails the loss of mucus, it also means that water is also lost, making them dependent on damp habitats for their survival. Snails see through eyes at the tip of two unequally long retractable tentacles and breathe through a respiratory pore. Snails are more often seen above ground in foliage. They retract into their shells, which are nearly waterproof, during the day, particularly when it is dry. In very hot weather sails will climb up to the tops of plants where it is cooler. If the weather is particularly dry or cold and the snail remains inactive it secretes a layer of mucus called the epiphragm across the opening to its shell and can survive like this for several months. Like slugs they prefer gardens where there is abundant lush vegetation and dead organic matter where they lay their eggs. One method of controlling snails is to expose their eggs to cold dry air by removing such rank vegetation. Snails possess the same type of rasping tongue covered with minute teeth as slugs and so cause similar damage to leaves and flowers. Generally snails are less important than slugs but they can be damaging on vigorously growing plants. It is advisable not to use too much organic fertiliser, manure and compost if there is a snail problem. However, such plant material together with sacking can be used to trap snails as well as slugs. Snails are also trapped by liquid baits such as stale beer or milk. Like slugs, snails are difficult to control, so efforts to eliminate them from a garden are doomed to fail where infestations are heavy. As for slugs, the most common method of checking their predation has been to use pellets of a bran bait poisoned with metaldehyde or methiocarb.

There are many species of snails – over 100 in southern England – which can be identified by details of shell shape and colour, but most are so small that they are easily overlooked. The snail that is most commonly found in gardens is *Helix aspersa*, the garden snail, which has a dark grey body and a banded pale brown shell about 4cm in diameter. They tend to be found wherever there are overgrown herbaceous borders, on dry stone walls, rockeries and among shrubs. They can be picked off by hand and dropped into a concentrated salt solution or hot water but are difficult to find during the winter when they hibernate in any available cavity in walls or beneath stones or rank vegetation.

There are several other species of snails that are important including the strawberry snail, *Trichia striolata* and the banded snails, *Cepacea nemoralis* and *C. hortensis*. The strawberry snail is only 1cm in diameter

and its shell colour is somewhat variable, ranging from dark grey to reddish brown. The banded snails have shells that vary in colour combinations from white, yellow, grey to pinkish usually with a darker band. They generally cause less damage than the garden or strawberry snails.

Sooty moulds

Several darkly pigmented non-pathogenic moulds, including some *Cladosporium* species are found as dark deposits on the upper surface of leaves resembling soot. These plants are either infested with an insect pest that feeds on sap or are near a plant that is host to these pests. The reason for this is that whenever aphids, adelgids, whiteflies, scale insects and mealy bugs feed on plant sap they excrete honeydew, a sticky sugary liquid, drops of which spatter onto the foliage and act as a food source for the sooty moulds. As the drops of honeydew form a continuous film over the leaves, the sooty moulds create thicker felt-like masses that disfigure the leaves. Although these fungi do not directly attack the plant, these black sheets of fungi prevent light from reaching the leaves. Although it is possible to wash off the fungi for lasting control it is essential to eliminate the pests by the use of appropriate insecticides. As well as foliage, sooty moulds are found on the glass and paintwork of greenhouses and also on cars if they are parked under an infested tree. In both cases it is essential to wash off the sooty mould before permanent damage is done.

Symphylids

The symphylids are related to the millepedes and centipedes and the only pest of any importance is the glasshouse symphylid, white insect or glasshouse centipede (*Scutigerella immaculata*). Although the main damage caused by the symphylids is to the root system of plants causing them to stunt, wilt or even die, if root rot pathogens can enter wounds, leaves that touch the soil are also often eaten as well. Insecticides usually do not make contact with symphylids as these pests survive deep in the soil. It therefore is usually much easier to avoid allowing them to become established rather than attempting to eradicate them, so it is prudent not to bring in plants or soil from areas known to be already infested.

Thrips

Thrips are the tiny slender insects (*Thysanoptera*) usually a couple of millimetres long, commonly known as 'thunder bugs' or 'thunder flies'.

Adult thrips have hairy fringes to their narrow wings. The wingless immature stages resemble the parents but are paler in colour. Thrips may be so tiny that you may not be able to see them clearly as they wriggle through the flowers that they attack, but in some years they can really cause serious damage. They infest flowering fruit trees, such as crab apples, and many other flowering plants, like clematis, broom and roses, as well as a number of house plants including some ornamental ferns, are often badly affected.

Many different species of thrips attack garden plants. Several kinds are very common, but you may not notice that they are present. Thrips breed rapidly during the hottest summer months. They are those dark, almost insignificant, irritatingly minute and thin, wriggling insects that spoil outdoor activities on sweltering days in mid-summer. At this time swarms of the adults crawl everywhere, even over human skin, but unlike cat fleas, thrips do not jump and are much thinner, with two pairs of very slender wings bordered by delicate elongated hairs. Thrips vary in colour from amber to deep brownish black, depending on the species (there are likely to be a total of between one and 200 species in one area). Males are often quite rare in many common species of thrips; the females lay their eggs in plant tissues without the need to breed with males. Immature and adult thrips often spend the winter in the soil and leaf litter which protect them from the extremes of cold.

Several different species of thrips occur on one or more type of garden plant. Gladiolus thrips can soon disfigure gladiolus plants. The thrips produce numerous pale dots mark wherever they chew the flowers and can also be found under the bases of leaves. July and August, when the yellowish larvae as well as the brown adults are present, is the time when the most damage to flowers occurs. Initially these just look very dingy; afterwards during severe attacks some flower buds may die and then plants will often wilt.

Gladiolus thrips can conceal themselves in the scales of corms that have been stored over winter. These thrips can easily be controlled by a choice of one of several appropriate insecticides. Some thrips survive underground until the spring.

When leaves and flowers show light speckled flecking and silvering it is likely that they have been attacked by swarms of the yellow or brown larvae and adults of onion thrips. These can spread tomato spotted wilt virus. Many other plants are affected, but arums, begonias, cabbages, carnations, chrysanthemums, cinerarias, cyclamens, dahlias, gerberas, gloxinias, orchids, peas, sweet peas, pelargoniums and tomatoes are most often attacked.

Another culprit may be the glasshouse thrip. This is found on indoor azaleas, citrus, ferns, fuchsias, orchids and zantedeschias, but despite its name, it is generally less frequent in glasshouses than the onion thrip which

Left: **Garden snail travelling up stem.**

Above: **Western flower thrip.**

is found almost everywhere. As they feed, glasshouse thrips, unlike onion thrips, often stain the plants that they are on with reddish brown droplets. They also have brownish bodies with a pale yellow tip to the tail end.

Privet thrips, which have sets of darker and paler wings, attack privet and lilac bushes are often turned silvery, the growth of new plants is very severely repressed and serious infestations can bring on early leaf-fall. Rose thrips attack the leaves of roses and some other plants which turn silvery, their flowers at first become flecked and streaked, then darken and rot. Some rose thrips continue to breed on indoor plants which can be severely damaged, but outdoors they overwinter in cracks in brickwork, woodwork and canes.

Contact insecticide sprays, dusts or fumigants should be applied as soon as thrips or their symptoms are seen, and repeated after two to three weeks if new growth is still infested. Thrip numbers are reduced following cold, wet conditions, so in order to keep plants free of thrips without using chemicals, they should be watered regularly to keep the atmosphere as cool and humid as possible. Poor growing conditions resulting from inadequate watering after sweltering droughts encourage infestations.

Once they have become well established it is almost impossible to control serious outbreaks of thrips completely without using an appropriate insecticide, so all insecticide treatments should be applied as soon as the symptoms of thrips are first discovered. You should first repeat the application of insecticide if signs of thrip injury persist; however, if you still continue to notice any fresh symptoms, it would be wise to consider trying a totally different type of insecticide treatment.

Bugs feed on plants through their needle-like stylets which are held under the body when not being used. The two main types of bugs that are encountered on foliage are capsid bugs (see p. 57) and lace bugs. The rhododendron bug (*Stephanitis rhododendri*) is the only lace bug which is common enough to cause serious damage. Although *Rhododendron ponticum*, its hybrids and some other species are not affected, it attacks many other rhododendron species and hybrids. The adult rhododendron bugs are about 4mm long and black covered with wide lacy cream-coloured wings, although they rarely fly far. They are first seen around midsummer and lay their eggs on the underside of rhododendron leaves under drops of excrement which dry to form a protective surface over the eggs until the autumn. The slow moving nymphs are yellowish with brown spines and congregate in small groups which usually stay on the leaf where the eggs were laid. The adults also rarely move off the lower surface of the same rhododendron leaf on which they hatched and fed until they are ready to lay their eggs on new foliage nearby. Several insecticides can control lace bugs if applied as soon as the infestation is seen.

Viruses

Viruses possess some of the attributes of living organisms but also show some of the other properties associated with non-living matter. They lack the ability to reproduce outside the host and most become inactivated by a even short spell outside it. They depend on the host to provide material for them to replicate themselves. Viruses are so small that they cannot be seen with a light microscope unless they form crystals. They are often spread through vegetative propagation and so infected plants with no or few symptoms other than stunting are often regarded as normal, it is only when the plants are freed of the virus that their true effects are recognised. Many viruses can remain latent within the host without revealing any symptoms. Traditionally viruses are detected by grafting a piece of infected plant onto a healthy one. This test can detect even those viruses that are latent but is not sufficient evidence that the infection is caused by a virus as some other pathogens can be transmitted in this way. When viruses are purified they have been found to consist of one or more pieces of nucleic acid, nearly always RNA, coated with proteins. This allows professional plant pathologists to devise techniques to recognise very small changes in the structure of the proteins and nucleic acid between different viruses even in very low numbers. Unfortunately, neither of these very accurate techniques is available to gardeners yet.

In the garden the only evidence of the virus available is the symptom which means that the virus has already caused damage. Also many viruses cause rather unclear symptoms. However, in some cases there is a more distinct symptom. Where the leaves are variegated or show two or more colour types, one of the most common types of symptom is mosaic, where there is a mixture of green and yellow tissues usually limited by the leaf veins. Where the leaf has longitudinal veins the symptom is a streak. These symptoms are related to vein banding where the leaf veins are yellow and vein clearing where the veins are clear. Another similar symptom to mosaic is mottle, here the yellowing is more clearly defined and rounded. If there are smaller areas of mottle, they are termed blotches, flecks and spots. If there are rings of lighter tissue on an otherwise green leaf this is known a ringspot. Some plants can show patterns of yellow lines, these line patterns may resemble shapes such as oak leaves. Where the green tissue is completely transformed to yellow, this is known as yellowing. Other symptom types are related to malformation include leaf curl and leaf roll. Some viruses kill part of the plant resulting in necrosis. A number of viruses affect the pigmentation of flowers, where new patterns of colour change appear due to the action of a flower-breaking virus. Combinations of these and other symptoms represent the syndrome of each viral disease.

Although some viral diseases are suppressed if the insect, often aphid, vectors are controlled this is not really possible under garden conditions. Some garden plants are actually highly valued because of the variegation. Some of these plants are virus infected although in others the variegation is genetically controlled.

Weevils

Apart from the vine weevils which are dealt with elsewhere, several other weevils are troublesome in the garden. Several leaf weevils (*Phyllobius spp.*) can be severely damaging as adults to a number of trees and shrubs including ornamental cherry and rhododendron. The leaf weevils eat large holes out of the leaves and petals. Generally these leaf weevils are up to 10mm long and many are brown, as, for example, the brown leaf weevil (*Phyllobius oblongus*) but some are metallic, such as the common leaf weevil (*P. pyri*) which is greenish-bronze and the silver green leaf weevil (*P. argentatus*). Although insecticides are often used, these are not necessary if the tree or shrub that is infested is small enough to be shaken over a sheet so the weevils caught can be destroyed.

Clay-coloured weevils (*Otiorhynchus singularis*) are similar in size, but their brown background colour is speckled with lighter marks; they are wingless, nocturnal and found from early spring. Clay-coloured weevils attack the leaves and bark of many fruit trees as well as ornamental trees and shrubs but they remain in hiding in the soil and under plant debris by day. Among their favourite hosts are apple, buddleia, cherry, clematis, primula, rhododendron, rose, wisteria and yew. As well as eating holes through the leaves, destroying buds especially on grafts, cutting off leaves and flowers by biting through the leaf stalks or petioles and flower stalks, they also notch the leaf margin in the same way as the vine weevil. In addition, they chew the bark from shoots, many of which die if they are girdled. The larvae that hatch from the eggs laid in the soil feed on the roots of plants during the summer and can damage some smaller plants. The adults and larvae of the figwort weevils (*Cionus scrophulariae* and *C. hortulanus*) attack the leaves of buddleia, phygelius and verbascum. The larvae are covered with a yellow-green slime as they strip one surface of the leaf, leaving the other to burst open afterwards. The rather square-shaped adult figwort weevils are grey with black spots and about 4mm long. These are generally found feeding on the young leaves, their petioles and the flowers and their stalks. The nut leaf weevil (*Strophosomus melanogrammus*) is sometimes a nuisance on rhododendron bushes growing near woodland, as the 6mm long grey-brown adults gnaw the upper surfaces and margins of the leaves and stem surface. Most leaf weevils respond to insecticide sprays.

White blister

Ornamental crucifers such as honesty are frequently infected by *Albugo candida* and often the infection is by contact with the weed shepherd's purse. The initial symptom is a raised shining white blister which consists of a raised epidermis above a sporulating pustule of the fungus, but at later stages the whole plant may become distorted. The sporangia of the fungus are primarily distributed by wind and rain to new hosts, there they germinate to produce zoospores which enter the stomata. Although fungicide sprays can help, it is wise to remove all diseased plants and weeds as soon as possible.

Whiteflies

Although whiteflies are tiny, they are rarely confused with other small insects because of their distinctive behaviour and appearance. The adults resemble minute white moths just over 1mm long, but they are not closely related to them and their wings and bodies are covered with powdery wax, not scales as in the *Lepidoptera.* Large numbers of whitefly usually remain out of sight on the undersides of leaves and growing tips where they suck the sap of the host unless they are disturbed, in which case they flutter around in a characteristic and very noticeable way. The small, flat, oval nymphs, often called scales, also inhabit the undersides of the leaves where they too suck sap but are well camouflaged as they are colourless and virtually transparent until they pupate into thicker, white, wax-covered pupae. The presence of sooty mould fungi is associated with whitefly as they grow on the honeydew that the larvae excrete in tiny sticky droplets over the leaves. There are several species of whitefly outdoors – during the summer these may include the glasshouse whitefly (*Trialeurodes vaporariorum*) that is a major pest of nearly all glasshouse plants. The glasshouse whitefly is particularly damaging to abutilon, some begonia varieties, calceolaria, chrysanthemum, cineraria, dahlia, freesia, fuchsia, heliotrope, primula, salvia and pelargonium. The main symptoms are yellowing and mottling of the foliage followed by stunting, wilting and death if the plants are heavily infested. Sooty moulds and specks of honeydew make the plants unsightly, but by the time that these symptoms are apparent the plants will already be well colonised by several generations of whitefly. The glasshouse whitefly generally reproduces by parthenogenesis, from eggs laid under the leaves in circular groups, providing there is an adequate supply of food plants. If these are absent the whitefly hibernate over winter on any plants that are handy until the following season. Although the pale green nymphs initially roam around, once they find a suitable place to feed they stay there until

Top left: **Yellow stripe virus symptoms.**
Left: **Vine weevil on begonia stem.**
Top: **White blister** ***Above:*** **Whitefly**

they are gradually coated by a white wax that they secrete as they grow into adults – over three to four weeks in summer but longer in winter. Infested plants will usually have representatives of all these generations present at the same time.

Although some insecticides give some control if used at any early stage of infestation, cheap and effective biological control can be achieved by releasing the tiny parasitic wasp (*Encarsia formosa*) which exterminates the nymphs and adults but is harmless otherwise and does not sting people or pets. This wasp, which is about a quarter of the size of a fruit fly, lays its eggs in up to 150 of the larger, more mature whitefly nymph scales. It prefers a temperature between 18 and 30°C and is inhibited by sooty mould so any leaves that are affected by whitefly are best removed by hand before treatment. Once the wasp larvae hatch they start to devour the whitefly larvae from within. These parasitised scales turn black and eventually produce more *Encarsia* wasps instead of whiteflies. *Encarsia* wasps are available by mail order in the form of parasitised whitefly scales that are pasted onto cards and are hung in the glasshouse as soon as they arrive. Two batches of cards are usually supplied with a gap of a fortnight between them. If necessary the cards of parasitised whitefly scales can be held in the refrigerator for a few hours, although this will reduce the numbers of wasps that emerge. It is difficult to tell when the wasps have emerged as the scales look unchanged unless closely inspected. If large numbers of whitefly flutter out if plants are shaken, an appropriate insecticide is sometimes applied before the wasps are released. However, the wrong insecticides or sticky traps will reduce the effectiveness of the wasps drastically and should be avoided. It is sensible to destroy any weeds such as chickweed, nettle and sowthistle that become infested near glasshouses and conservatories.

Among the whiteflies that are found outdoors are the rhododendron whitefly (*Dialeurodes chittendeni*), the azalea whitefly (*Pealius azaleae*) which is found on evergreen azaleas, the honeysuckle whitefly (*Aleyrodes fragariae = lonicerae*) on honeysuckle and snowberry and the viburnum whitefly (*Aleurotrchelus jelinekii*) found on *V. tinus*. The rhododendron whitefly can be a pest on several choice rhododendron species with relatively thin cuticles, such as *R. campylocarpum* and *R. catawbiense*, as well as the common *R. ponticum*. Other rhododendron species that have woolly or scaly undersides to their leaves are rarely affected, probably because the adults find it too inconvenient to lay their single eggs on these leaves. Attacks by the rhododendron whitefly are revealed by the adult whitefly which flutter when the leaves are brushed aside in June/July, as well as presence of the empty moulted skins of 'pupae'on the underside of the leaves from mid-April, but the greenish nymphs, which also live there from mid-July and then through the winter, are too well camouflaged to be seen easily. After a time rhododendron bushes that are heavily attacked by whitefly become mottled yellow but do not show the brown discoloration on the underside of the leaves associated with the rhododendron bug. The leaves are also spotted with honeydew and the secondary sooty moulds. Heavily infested branches or plants are best pruned out and burnt. In less serious attacks, several insecticides are effective as sprays against outdoor whiteflies, but fortnightly applications may prove necessary.

Woodlice

Although woodlice or slaters are land-living relations of other crustaceans which are found in water, including the shrimps, crabs and lobsters, they are generally found in damp places from which they foray out at night to feed. Woodlice eat a wide range of decomposing organic matter including the roots and stems of plants, especially seedlings. As a result the cotyledons and young leaves can be severely damaged. Woodlice are often most troublesome where there are stones and woodwork under which they can hide, for example, in corners of glasshouses, cold frames and near rockeries, particularly if decaying plant material is left lying around. There are several species of woodlice including *Oniscus asellus*, *Porcello scaber* and the pillbug, *Armadillidium vulgare*, which can roll itself up into a ball. In all of them, hard grey-brown plates cover seven pairs of walking legs, and the head has two pairs of antennae, with one pair much longer than the other. Female woodlice carry up to two broods of eggs and, after they hatch, carry the white young in a pouch under their thorax. The most effective way of controlling woodlice is to clear up all decaying plant material, and decaying woodwork especially in glasshouses and cold frames where young plants are propagated. If infestations persist some insecticides are effective. Rockery stones should be firmly seated in the soil and should not provide hiding places suitable for woodlice.

***Above*: Grey garden woodlouse on plant stems.**

(iv) Roots

Chafer beetles

There are four common species of chafer beetles whose 'C'-shaped larvae can be found damaging the roots of garden plants. The largest and probably the most frequent of these is the May bug or the cockchafer (*Melolontha melolontha*) found in May/early July. The adult is about 25mm long with straw-coloured wing cases, a black thorax and head with comb-like antennae. The larva can reach almost 40mm in length. The much smaller garden chafer (*Phyllopertha horticola*) is also common in May/June, and unlike the others the adult is active during daytime. Compared to the cockchafer, the adult garden chafer is about half as big, while the larva is just over a third of the size. The adult summer chafer (*Amphimallon solstitialis*) found in June/July is slightly larger than the garden chafer, but the larva can reach 32mm long. The rose chafer (*Cetonia aurata*) is 19mm long and has an unmistakable metallic greenish sheen on its wing cases which are marked with white flecks. This is the least widespread of the chafers but can be common in a few localities in May/June. The larva is about the same size as that of the summer chafer but can be distinguished from it and the other chafer larvae by the presence of transverse rows of reddish hairs along its body.

Adult cockchafers can be found flying to lights at night when they feed on the leaves of a variety of plants, both herbaceous and woody. In addition, they feed on the buds and flowers of roses and gnaw into the skins of developing apple and pear fruits. During the day resting cockchafers can be found among foliage. After mating has taken place the female cockchafer burrows into the soil to root level and lays batches of 12 to 30 eggs which hatch after a few weeks into larvae. These larvae do not cause significant damage until the following year when the larvae increase in size and feed until the next summer before pupating. About six weeks after pupation the adults leave the pupa but remain below ground in the earthen cell until the subsequent May/June – the complete life cycle therefore lasts four years. The other chafers have similar but shorter life cycles, particularly the garden chafer which is complete in one year. The larvae of the rose and summer chafers feed for two to three years.

The adult chafers can be shaken from trees and shrubs and destroyed or their daytime haunts can be sprayed with a suitable insecticide, but while these methods reduce the population of roosting adult beetles, at night a new population will fly in and lay more eggs. For this reason it is more sensible to concentrate on eradicating the larvae. This is most easily done by cultivating the soil, but where this is not possible, as on established lawns, an approved insecticide may be applied to the soil and worked in. On lawns the insecticide can be applied with sand or fine soil to make sure it is evenly spread. If heavy chafer damage is likely an insecticide can be incorporated into the soil when trees and shrubs are planted.

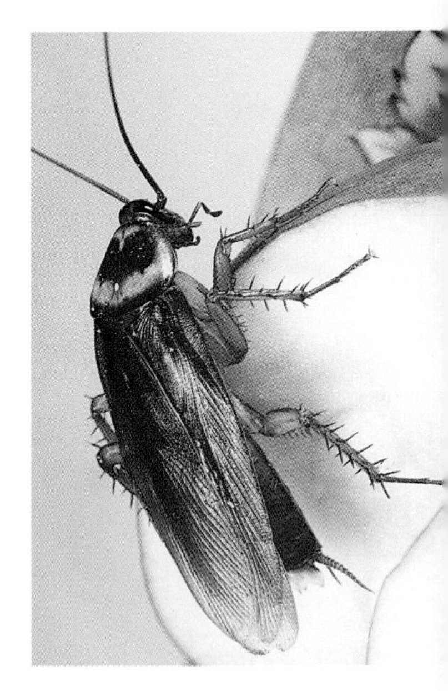

Cockroaches

These fast moving insects with spindly legs and long antennae can sometimes be very damaging to plants in heated glasshouses. There are two common species, the blackish oriental cockroach (*Blatta orientalis*) and the larger reddish brown American cockroach (*Periplaneta americana*). Both species are nocturnal but may be found hiding during the day, especially around hot water pipes. During the night they feed on any plants that they can reach including seeds, seedlings, foliage, stems and aerial roots. Their favourite plants include the flowers of chrysanthemum, cineraria and orchids. Under suitably warm conditions cockroaches can breed around the year. Female cockroaches can often be seen carrying a purse-shaped capsule containing a batch of 16 eggs until they find a suitable crack or crevice in which to hide them. After one to three months, depending on the temperature, the eggs hatch, and within 13 to 16 months the cycle is repeated. Cockroaches are very difficult to eradicate as the breeding season is so prolonged and the eggs take so long to hatch that new generations are continually being produced. Cockroaches are also often resistant to the insecticides placed on their favourite haunts, so several applications may be necessary. Another way of controlling cockroaches is to trap them in jars buried up to their tops in soil, using a bait such as beer, treacle or ripe bananas. These baits may contain insecticide. It is important to remove any plant debris that can either provide hiding places or food for cockroaches.

***Left*: Cockchafer grub.**
***Top*: American cockroach.**
***Above*: House cricket.**

Crickets

The house cricket (*Acheta domesticus*) can damage plants by gnawing at the aerial roots and stems of a number of glasshouse plants including orchids at ground level and may also damage the foliage or flowers. It is able to jump quite long distances using its powerful back legs. Usually these crickets move around and feed at night but hide during the day. Their presence is usually revealed at night when they chirp by rubbing their front wings together. The mole cricket (*Gryllotalpa gryllotalpa*) produces an even louder call and does even more damage to the roots and stems of plants at or below ground level, particularly in glasshouses where it burrows into the soil. However, mole crickets are rare in Britain and so do not pose a serious problem in most gardens. As house crickets need shelter during the day it pays to remove any plant debris under which they could hide. Suitable approved insecticides can be applied to the soil where crickets are present, and as they are attracted to hot water pipes these areas should be especially thoroughly treated. Some baits incorporating insecticide may also be available.

Cutworms

The caterpillars of several moths are known as cutworms. Although several species involved are the turnip moth (*Agrotis segetum*), the heart and dart (*A. exlamationis*) and the large yellow underwing moth (*Noctua pronuba*). The adults of all of these are actively laying eggs on the leaves of a variety of plants in June and July. Once the eggs hatch a couple of weeks later the cutworms feed voraciously attacking many types of plant at ground level during the night but remain inactive during the day hidden curled up in the soil or rank vegetation. Among the plants that are most frequently attacked are asters, chrysanthemum, marigold, petunia and phlox, but many other succulent species are susceptible, including many weeds. The lower leaves and the stems are eaten so that the plants often die after they have fallen over. Cutworms also burrow into rhizomes and tubers. Some of this first generation of caterpillars pupate quickly and the moths emerge in late August/September and produce a second generation of cutworms. Nonetheless the majority of the first-generation caterpillars develop less rapidly and together with those from the second generation overwinter as larvae which pupate in the spring ready to repeat the life cycle again the following summer.

Care should be taken to remove potential weed hosts near garden plants. Suitable insecticides can be applied to the soil around susceptible host plants. Sometimes bran-based baits are used. Vegetation should not be allowed to become too rank and so provide daytime refuges for cutworms.

Damping-off

These soil-borne fungi usually include a number of common species of Omycetes such as species of *Phytophthora* and *Pythium*. However, these plants may often be attacked simultaneously by many other completely unrelated fungi, like *Fusarium* species, *Rhizoctonia solani* and *Thielaviopsis basicola* and *Aphanomyces euteiches*. Although an individual pathogen may predominate in a particular sample of soil, it is usually also heavily infected by many other pathogens to some extent. For this reason, it is usually quite hopeless to try to link any distinct symptom to a specific named pathogen as they lack any characteristic symptoms. In fact the effects of several pathogens are so similar that for practical purposes they are often grouped together even though their biological differences as well as the principal control treatments for them are quite different.

Nonetheless, the general behaviour of the unspecialised pathogens that are repeatedly associated with damping-off, root and foot rots have several fundamental features in common. With a few prominent exceptions, these fungi are ubiquitous in all soils as a mycelium that originally germinated from a spore. In practice, many soil-borne fungi develop from a resting spore or another resistant structure such as a sclerotium. These active hyphae are generally also able to live saprophytically on dead plants. Most of these can infect undamaged roots but many others need tiny wounds through which to infect plants. The latter are usually rather more pronounced on plants growing under a stress, such as drought, waterlogging or particularly poor light.

Groups of young seedlings frequently become weakened and die just after the first few leaves have formed. Initially seedlings topple over around several points that often then enlarge into round patches over the soil. When you pull out these seedlings their roots are usually missing or often extremely reduced to no more than a stub. These are the commonest diseases to affect the seedlings of a very wide range of bedding plants, especially fast growing species such as alyssum, antirrhinum, nemesia, penstemon, petunia, salvia, stocks and tagetes. Seedlings of herbaceous plants are regularly afflicted and rarely escape some losses in seed boxes or outdoor beds. Even tree seedlings are prone to damping-off, which can result in the death of at least some of the seedlings, particularly if grey mould is also present. Few of the fungicides and none of the fumigants applied for soil treatment in commercial operations are available for use by gardeners who must therefore rely on good hygiene after buying sterilised composts or using garden soil that they have steamed or baked in an oven for an hour or microwaved for five minutes themselves. Contaminated compost and diseased plants must be disposed of without affecting clean plants and pots. Apparently healthy seedlings from a bed or box containing some diseased individuals must not be used as they may themselves already be imperceptibly affected and thus can spread the disease. Seed boxes must be disinfected before being filled with sterilised compost. Larger plants also have stem lesions at or about soil level and their roots may rot away completely. Often the plants are the survivors of damping-off as seedlings.

Varietal resistance is not clearly substantiated for the seedlings of any of these crops, but some plant varieties are often more easily differentiated

when mature. However, a few plants actually become more susceptible to disease as they mature – these are principally plants with swollen storage organs, such as bulbs and corms. Here, although the main effects are on the roots or stem base, the first evidence of this underground damage is frequently shown by the leaves which are very much smaller than on healthy plants. Frequently these plants will abruptly turn yellow and wilt. flowering and fruiting may also suddenly decline. So if you think you can see this happening, you could try pulling up a few plants to see if the root systems are noticeably weak and blackened or even putrefied. Rots may occur near the root apex or in other distinctive zones.

Dead or dying plants should be removed and destroyed by burning. Fungicide-treated seeds, a well prepared seedbed and other good plant husbandry practices that improve drainage or balance the nutrient supply and acidity are usually the only possible means to ensure reasonably successful germination in garden soil.

Provided there are no susceptible weeds, rotations are only of some use if at least four or five years have elapsed between highly susceptible plants. However, without planting a few seeds, it is impossible to check whether this break has been effective. Instead, it may be quicker to dig in plenty of a non-susceptible crop. This debris appears to activate the population of antagonistic micro-organisms in the soil. Sometimes a special crop of brassica plants is grown for this purpose as a 'green manure'.

However, great attention must also be paid to protecting your supply of water from all sources of contamination. Rainwater butts should be cleaned out periodically to remove fallen leaves and other debris that could support some of the more regularly aquatic Oomycetes, such as species of *Phytophthora* and *Pythium*.

Eelworms

Eelworms or nematodes can cause substantial damage to many common garden plants and are particularly serious as they are pernicious and are so difficult to eradicate from soil once it has been infested. There are no chemicals that can be used by gardeners, and even professional growers have few products that are not extremely toxic or damaging to the environment. It is much better to avoid contaminating clean soil as it can remain affected for many years, making it difficult to grow some plants successfully. This happens because nematodes can exist in a dormant state for several years without feeding. Since chemical control is not feasible, biological control with bacteria is being investigated, but at present the best way to eradicate nematodes from garden soil is to keep the infested soil free from those weeds and cultivated plants that act as hosts. In this way the nematodes are denied a food source and will eventually be starved out until the population can be reduced sufficiently for susceptible crops to be grown again, even though this can prove to be a rather long process. In this case it is best to make sure that the susceptible host crop is rotated with crops that are known to be resistant. Even this system of keeping land fallow and practising rotations is not infallible. Vigilance is essential, as several of the weed hosts and some of the cultivated ones do not show clear symptoms of infestation but act nonetheless as reservoirs of infection. Most nematodes are wriggling filamentous organisms less than 1mm long which are normally only clearly seen under the microscope where they can be observed to be highly active, translucent, smooth and worm-like. The hind end is more pointed than the head. Some species produce cysts which are visible to the naked eye. As they move through the soil from plant to plant their progress can be affected by the type of soil that they encounter. Nematodes tend not to infest fine-particled clay soils which are easily compacted, dry soils, nor waterlogged soils which are deficient in oxygen.

Nematodes generally feed through a hollow mouth-spear; this is used to stab into plant cells and suck out the contents. After a time as a result of repeated feeding on neighbouring cells, the eelworm creates patches of dead cells which may then decay. During the process of feeding some nematodes inject saliva into the cells to digest them. This often results in the enlargement of the cells which swell the affected tissue, thus distorting the roots, stems or leaves. In most cases, although only relatively small fresh areas of plants are attacked each year, these patches of infestation can increase over the years. Nematodes breed throughout the year by producing eggs that pass through four juvenile larval stages before developing into adults.

There are a number of very destructive nematodes which attack plants below the soil. The potato tuber eelworm (*Ditylenchus destructor*) attacks a wide range of plants in addition to potato, including many bulbs, corms, *Dahlia* root tubers and lilac as well as many common weeds such as some mints and sowthistles. Although the symptoms vary, usually the affected plants have dark rotting areas but if tubers are infested soft dry patches develop around these rots. In *Iris* thin, vertical, black lines appear in the scales which merge to rot the bulb. Prevention of spread is easier than cure for amateur gardeners; growers use fumigants to sterilise the soil and hot water dips for bulbs, but this is rather exacting as too high a temperature can damage the bulb and too low will have no effect.

The stem eelworm (*Ditylenchus dipsaci*) has an even wider host range than its relative, the potato tuber eelworm. Plants particularly affected include Narcissus, tulip and other bulbs – *Hydrangea, Oenothera, Phlox* and

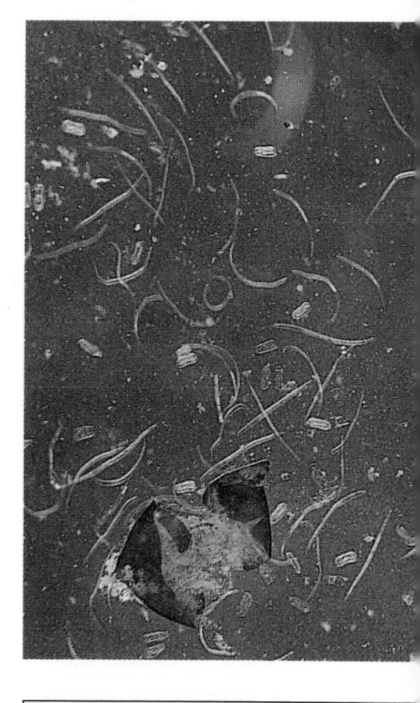

Left: **Damping off of seedlings.**

Above: **Cyst broken open to reveal eelworms.**

Primula. It is more serious in its effects and susceptible plants often decline rapidly, but many others show few symptoms and act as the reservoirs of infection for the former. There are several races, each of which attack a limited range of particular plant hosts. The stem eelworm enters the bulb, stem or leaves from the soil through natural openings such as the stomata and lenticels or small wounds. Once inside the nematodes disrupt the stem tissues and bulbs to form spongy patches as well as dark swollen areas which stunt the plant. These lesions rot and spread until the plant collapses, but the roots are rarely affected. Again, prevention is more achievable than control as similar methods are needed to those used for potato tuber eelworm.

In contrast with these, the cyst eelworms (*Heterodera* spp.) and their relatives, the root knot eelworms (*Meloidogyne* ssp.), are found only on roots. The second-stage larvae of both penetrate through the root until they reach its centre where they feed and inject saliva into cells which expand around their heads to feed the developing eelworms. The fuctioning of the plant deteriorates as these giant cells disrupt the usual operation of the roots.

Some other eelworms can move freely from one root to another to feed on the root surface or like the root-lesion eelworms to form breeding colonies unlike the more permanent root parasites, the cyst and root-knot eelworms. Root-lesion eelworms (*Pratylenchus* spp.) attack roots, bulbs, corms, tubers and rhizomes of several plants including clematis, delphinium, helleborus and thalictrum to produce slit-like lesions that allow bacteria to rot the underlying tissues and extend the lesion along the length of the root. They are frequently found in sandy soil.

Pratylenchus penetrans causes root rot of narcissus and *P. vulnus* attacks the roots of glasshouse roses. These nematodes can be controlled when bulbs and corms are dried as they are easily desiccated and are hence rarely transmitted to clean soil. It is more difficult to prevent their spread on rooted plants so it is sensible to take cuttings of these and root them in sterilised compost. African and French marigolds (*Tagetes* spp.) are antagonistic to *Pratylenchus* spp. so it is worth digging these into areas which are affected before planting with susceptible hosts. Many other nematodes cause damage by feeding on the surface of plant roots – these include the stubby-root eelworms (*Trichodorus* spp.), needle eelworms (*Longidorus* spp.) and the dagger eelworm (*Xiphinema diversicaudatum*) which can transmit a number of virus diseases. It is possible to disinfest perennial plants which are not infected by virus by washing off all of the soil and replanting them in soil free from nematodes. However, these nematodes have so many different host plants that rotation rarely works.

Honey fungus root rot

In Britain honey fungus (*Armillaria mellea*) is the root disease most often responsible for causing death or decline in vigour of trees and shrubs. As well as a wide range of uninjured and apparently healthy, broad-leaved and coniferous trees and shrubs, herbaceous hosts such as iris, potato, strawberry, narcissus, bamboo and geranium can sometimes be attacked. Although frequent in established woodland, parks and gardens, and on fruit trees in orchards, it is less common where trees have been planted on non-forested land. Early stages of infection may be marked by increased cone, nut or fruit production and in conifers, a flow of resin from the base of the trunk. Meanwhile beneath the loosening bark, a network of blackened rhizomorphs develops and through the roots and lower trunk spreads the conspicuous whitish cream mycelium causing irregular patches of decay which finally become soft, wet and spongy, often bounded by black lines. Although young trees may be killed once the collar is girdled, the fungus may be confined to the roots or butts of larger trees for much longer causing them to become stag-headed or wind-thrown. After death, hardwood stumps can persist as sources of infection for decades. Although *Armillaria mellea* is widespread and common over much of the Northern Hemisphere, it was not distinguished from four or five of the related species found in Britain until fairly recently, so most of the earlier pathological literature on *A. mellea* needs careful interpretation. *A. mellea* can be separated from other pathogenic *Armillaria* species by its fruit bodies, but these are generally only found for a short time during the autumn. Large clumps of honey-coloured toadstools have a cap about 10cm across (at first covered with dark brown fibrillose scales which disappear as the cap becomes depressed with a striate margin) and a fibrous fleshy stripe (about 15cm long and about 7mm across) that is striate above the thick, white, membranous ring. The pale adnate or slightly decurrent gills bear abundant colourless basidiospores that appear white. When cooked the woolly flesh is sweet, but raw it is white and acrid, containing haemolytic substances, so its edibility has been questioned. Rhizomorphs – slender strings of mycelium covered in a dark rind – enable spread through the soil to enmesh and attach to the roots and collar. The dichotomously branched black rhizomorphs of *A. mellea* can be distinguished from those of weakly virulent *A. gallica* as these have longer brown monopodial rhizomorphs which are more elastic and bootlace-like. The sheets of white mycelium found under bark and in root decay can be distinguished by interfertility testing, immunology or nucleic acid hybridisation. Despite some evidence that *A. mellea* produces a phytotoxin, little is known about its mode of entry. Though possibly not as important as in other species, colonisation of

stumps by airborne basidiospores can initiate new foci in *A. mellea*, but afterwards spread is through the soil, by direct contact with infected roots or via rhizomorphs.

Although most infected woody hosts eventually succumb, comprehensive lists of apparently resistant and susceptible species have been based on field observations but the basis of susceptibility is inadequately understood. Physical barriers have not been very successful, so one should dig out and burn all dead wood. Chemical eradication by soil injection is occasionally practised in some countries prior to tree planting, and fumigants have been used in forestry to inject infected stumps. Research on chemical drenches and biological control continues; products based on cresylic acid are already approved by the MAFF.

Leatherjackets

Leatherjackets are also fly larvae: they are the larvae of crane flies, the rather delicate flies with easily detached legs, also known as daddy longlegs. Although there are a number of species, the only ones that cause serious damage are the greyish crane flies *Tipula paludosa*, *T. oleracea*, and the spotted crane fly (*Nephrotoma maculata*) which is yellow and black, and all have similar life cycles.

The adults emerge from pupae that force their way to the soil surface. *Tipula oleracea* and *Nephrotoma* maculata are mainly found flying to mate and lay eggs from May/August but *T. paludosa* is found in late August/early September. The eggs are laid into the soil often beneath thick layers of vegetation and soon hatch into larvae that feed on neighbouring plants providing the soil is moist enough. For this reason the greatest damage generally follows several months of wet weather following egg laying. When full grown, the legless leatherjacket maggots can range from 25mm to almost 40mm long. As they are greyish with a wrinkled leathery skin with no apparent head they can easily be overlooked as they are well camouflaged by the soil where they are found beneath the surface, feeding on roots and stems of plants. Mature plants often become stunted or wilt when the larger roots are destroyed, but seedlings are frequently completely devastated. Bulbs, corms and tubers are burrowed into and eaten. Occasionally when the weather is sufficiently humid the leatherjackets emerge onto the soil surface and will attack plant stems in a manner rather like cutworms. Lawn grass is often killed or thinned, leaving bare patches. The peak period for leatherjacket damage is the spring but losses may also extend to early summer in wet years or during mild winters. Leatherjackets are less common in dry seasons or dry sites as drought kills the eggs and young larvae unless they burrow into the deeper layers of the soil, but they are abundant in damp soils and wetter regions. Although several insecticides can be watered on and incorporated into the soil to control leatherjackets – preferably during warm damp weather when the leaterjackets are near the surface – simply making sure that the soil is kept well drained and clear of weedy vegetation by hoeing or other means of cultivation can also be effective. Another non-chemical method that can be tried involves trapping the larvae by thoroughly watering the infested soil then covering it with a tarpaulin. In the morning the leatherjackets that have emerged can be killed. If insecticides are used they can be incorporated into baits with bran or applied to lawns in sand or fine earth.

Millepedes

Millepedes are not insects but are related to them. Instead of the three pairs of legs found on the thorax in insects, the body of a millepede consists of numerous segments, each with a pair of legs set apart from the first of the three that form the thorax. There are no wings. Millepedes feed on rotting vegetation, but some species also feed on living plants where they are often attracted to lesions caused by disease pathogens or a range of other pests. These lesions are extended by millepedes, but they are also capable of tunnelling into soft areas of the roots and stems, seedlings are particularly vulnerable and seeds may also be destroyed by burrowing millepedes. This tunnelling can be severe during periods of drought as millepedes try to avoid dry conditions by obtaining moisture from plants. They breed during the spring and early summer when eggs are laid into the soil. The young millepedes that hatch have only three pairs of legs and only a few segments, but after they have completed seven moults these increase to the full complement.

There are two groups of harmful millepede species. The commonest group are the snake millepedes, which can be distinguished by smooth, hard, rounded bodies with slender short legs on 30 to 60 segments. Among the millepedes in this group is the spotted millepede (*Blaniulus guttulatus*) which is white with rows of reddish spots along its sides and is only 13mm long, and the black millipedes (*Cylindroiulus londinensis* and *Tachypodiulus niger*) which are somewhat larger and blackish in colour. Sometimes these millepedes are confused with centipedes which are beneficial predators that live in similar places, or even with wireworms, although the latter are beetle larvae which have only three pairs of legs. Most millepedes curl up when touched, unlike most centipedes which either actively run away or at least writhe.

Left: **Old honey fungus toadstools.**

Top: **Leatherjacket larva on lawn grass roots.**

Above: **Flat millepede on soil.**

The second group of millepedes which damage plants are the flat millepedes. flat millepedes have 19 to 20 flattened segments with a roughened cuticle which make them appear serrated. Compared to the snake millepedes, flat millepedes also have sturdier, more elongated legs. There are several common species including the 25mm dark brown *Polydesmus angustus*, 13mm cream-fawn brown *Brachydesmus superus* and the glasshouse millepede (*Oxidus gracilis*) which is reddish-brown and 19mm long. The glasshouse millepede is a tropical species that has been accidentally introduced into glasshouses.

Some insecticides can be dusted into seed drills or worked into the top layers of the soil before planting. If established plants are affected their roots should be drenched.

Phytophthora root rot

Phytophthora cinnamomi is now widespread on a number of trees (beech, apple, yew, limes, *Prunus* spp., planes, oak, walnut and conifers, particularly Lawson cypress) and many shrubs. It is especially common on rhododendron (including azaleas) and heathers, which wilt before they die. Many exotics are killed, such as pineapple, avocado, *Eucalyptus* and numerous other plants endemic to Australia. *Phytophthora cinnamomi* also causes ink disease of sweet chestnut in old coppiced woodland, sometimes in conjunction with *P. cambivora*. Infected roots develop a patchy necrosis that can extend up the trunk to the base of low-growing branches, causing the loss and stunting of foliage, often leading to partial or complete dieback of the tree. During hot dry summers, these symptoms have reached epidemic proportions in parts of England where Lawson cypress have been planted in heavy soils that become waterlogged in winter.

The pathogen (*Phytophthora cinnamomi*) probably originated in the tropics (possibly the islands off S.E. Asia and the South African Cape), but the A2 mating type, found in Britain is now a cosmopolitan soil-borne fungal pathogen (nonetheless, A1 still predominates in its original range). Although relatively unspecialised, some strain differences have been reported on several hosts.

Colonisation of the roots is by hyphae, produced initially by zoospores that swim and penetrate the root tissues after germinating from oospores that may have been present in the soil for up to ten years down to depths of over 700mm. Within the roots, the growth and penetration of the resultant mycelium results in extensive damage, killing the roots without decaying them. Wood-rotting organisms, such as *Armillaria* spp. often invade affected roots very rapidly, making it difficult to diagnose the coralloid hyphae of

P. cinnamomi. A relatively pure mycelium of *P. cinnamomi* that grows into a characteristic rosette pattern on agar, can be isolated from the advancing edge of the firm rot that develops within a few days around infested soil sealed in a cavity dug in a green apple.

Control of this disease is effected by transplanting only plants purchased from a reputable nursery or other area known to be disease-free. Some light soils can become suppressive if manured. Since plants of any sort with soil adhering to their roots may also spread the pathogen, select those grown in sterile compost in containers. Avoid any unhealthy browned cypresses or other trees. Resistant varieties such as Leyland cypress (x *Cupressocyparis leylandii*) can replace similar but susceptible species like Lawson cypress, permitting sustained freedom from disease. Do not transplant susceptible hosts into heavy soils, particularly those containing added peat or a similar medium, as these quickly become waterlogged, favouring the dissemination of zoospores. Avoid excessive watering or mulching. Do not leave a depression around a susceptible plant. Do not plant seed from wind-fallen fruit that may have become infected from contact with infested soil. Although a number of fungicides have MAFF Approval, these are only available for use by professionals not amateurs.

Root aphids

As well as the more familiar leaf aphids, there are a number of species that feed on the roots of plants. These aphids remain hidden underground, but their activity is noticeable when the plants that they infest become stunted and often wilt conspicuously in droughts long before healthy plants. Among the common root aphids are the artichoke tuber aphid (*Trama troglodytes*) which is found on chrysanthemums, helenium and sunflowers as well as a variety of weeds; the auricula root aphid (*Pemphigus auriculae*) on auricula and other *Primula* spp.; *Aploneura lentisci* and the dogwood aphid (*Anoecia corni*) which attacks the roots of grasses and the elder aphid (*Aphis sambuci*) which infests carnations, pinks and London pride. Some of these aphids – for example, *Pemphigus auriculae* – can easily be seen when plants are turned out of their pots as they produce a white waxy powder that covers the roots. The presence of other species of root aphids is often revealed by the profusion of active bands of attendant ants that protect the aphids while they herd and 'milk' them to excrete honeydew, often moving the aphids to the roots of fresh host plants. As with the aphids that attack foliage, root aphids are difficult to control completely. All aphids have an extraordinary capacity to reproduce themselves, so in an extremely short time after a few aphids have invaded a plant there is a very heavy and damaging infestation.

Above: **Phytophthora root rot on Lawson Cypress root**

If root aphids are suspected the plant should be dug up or turned out of its pot and examined for signs of infestation. If aphids are present the plants should be dipped into an aphicide solution before being repotted into clean compost or soil. Established plants may be drenched with an aphicide solution and the surrounding soil treated in the same way before any susceptible plants are planted there. Care should also be taken to eradicate any ants nests in the surrounding area as these may reintroduce fresh colonies of root aphids.

Root flies

There are a number of flies whose maggots attack the roots of ornamental plants. One of them, the cabbage root fly (*Erioischia brassicae*) attacks a range of cruciferous garden plants like *Aubretia*, stocks and wallfowers as well as the vegetable brassicas. Cabbage root flies can produce two or sometimes three generations in a single year starting in April/May. Eggs laid near susceptible host plants hatch into whitish maggots. The maggots are attracted to the host roots where they feed on the lateral roots then tunnel into the main roots eventually reaching the stem. The larvae pupate in the soil after feeding for about three weeks. The flies that form the second generation emerge from these pupae at the end of June/July and in late August their offspring will in turn have formed the next generation of flies. The adult fly resembles a dark grey house fly about 6mm long. Cabbage root flies are generally controlled by applying insecticide around the base of transplants a few days after planting. There is another method of protecting plants by placing a square of carpet or similar fabric around the stem of the susceptible plant to prevent the flies from laying their eggs adequately in the soil nearby.

Root mealybugs

Mealybugs are wingless, flattened soft, oval insects quite closely related to scale insects but differing from them in that neither the females nor the immature stages have scales but are protected by a covering of powdery or filamentous white wax. In contrast to scale insects, mealybugs are not stationary and are able to move as soon as they hatch from the egg mass which is laid in a waxy egg sac. Their bodies are fringed with stubby waxy structures sometimes with elongated ones at the tail end. Although many species of mealybugs attack the foliage, others live underground on the roots of a number of plants. These root mealybugs are frequently protected by ants which collect honeydew from them rather similar to the relationship between ants and root aphids. Root mealybugs are fairly widespread on grasses and a variety of wild plants most of which appear to be able to tolerate their feeding. However, root mealybugs such as the genus *Rhizoecus* can also be found on many glasshouse plants including abutilon, acacia, cacti and succulents, cassia, dianthus, dracaena, gardenia, pelargonium and stephanotis. Affected plants can appear to be wilting from drought, but when their roots are exposed the white waxy masses of the root mealybugs and their egg sacs can be seen. Once they have colonised the roots of the plants they are not very easily eradicated, but some success has been obtained by dipping the roots of infested plants in an appropriate approved insecticide and then repotting in clean soil. If plants are established then drenches of insecticide are worthwhile. Soil that is likely to be infested can be treated in a similar way. As ants are so closely associated with root mealybugs they should be eradicated from the area.

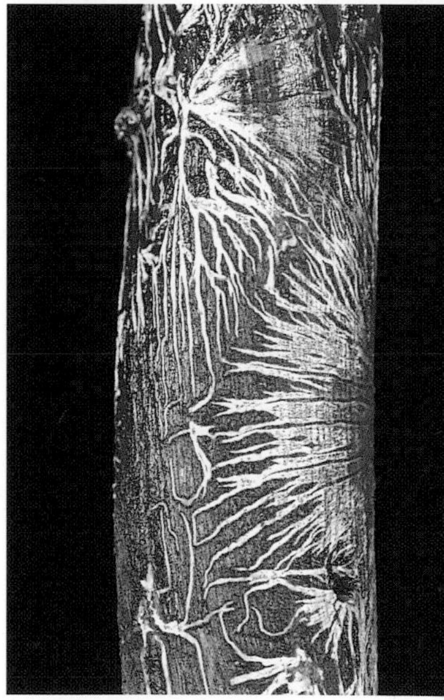

Top left: Root aphid on house plant roots.
Left: Root fly larva.
Above: Root mealy bug & white root rot.

Rosellinia white root rot

Rosellinia necatrix is widespread and frequent on a very wide variety of different garden plants, particularly in the milder, wetter, poorly drained regions such as the south west of England. It is especially common in warm wet seasons. Among the garden plants affected are acacia, almond trees, apple trees, begonia, cherry, chestnut, cotoneaster, cyclamen, elm, fig, genista, jasmine, irises, ixias, grapevines, mullbery, narcissus, oak, olive, pears, paeony, plum, privet, quince, ribes, rose, salix, strawberry, tulips, violet, walnut and zantedeschia; artichoke, blackberry, bean, beet, cereals and potatoes are also infected. When plants are attacked they are usually affected in groups and infection is confined to the roots. Infected plants often linger on with wilting and sparse foliage, growth is slow and eventually plants usually die. If bulbs are affected, the most obvious symptoms tend to be a brown rot under the blackened outer scales but the white root rot can be seen permeating the internal tissues when the bulb is cut in half.

When woody plants are affected the youngest roots are covered first with white mycelium, which develops into a dense, grey-green, cobweb-like mycelium which becomes even more darkened almost to black and coarser with time.

If the soil is particularly damp the mycelium will grow over its surface around infected hosts. Often minute black sclerotia are formed which can contaminate the soil for many years. Fruiting bodies, the perithecia, are formed under the soil in some countries. These release airborne ascospores. There are two major problems in the control of *Rosellinia necatrix* white rot; the difficulty of accurate diagnosis and the lack of suitable chemical control. Nonetheless, *Rosellinia necatrix* is distinguished from Honey Fungus Root Rot,

Armillaria mellea, by the absence of rhizomorphs. One of the most effective methods of cultural control is to dry out the fungus in the soil by digging or rotavating the soil during hot weather as it cannot withstand desiccation.

Sciarids

Most sciarid flies only feed on rotting vegetation and so are attracted to plant tissues which have already been injured and have started to decay as the result of disease. Some species of sciarid flies, notably *Bradysia paupera* can damage seedlings, cuttings and even established pot plants, particularly carnation, cyclamen, freesia and primula. Soil treated with dried blood fertilizer is particularly attractive. This sciarid is black, 3mm long and can complete its life cycle in less than seven weeks; sometimes this can be completed in only four weeks if conditions are sufficiently warm and otherwise favourable, so several generations of each stage of the lifecyle are usually present at the same time if breeding carries on continously. As well as feeding on the young roots, the larvae – which are 5mm long with a translucent whitish body and glistening black head – can tunnel into the main root and occasionally up inside the stem. As a result, the plants that are attacked become stunted and often wilt during warm dry weather. Many newly rooted cuttings and seedlings fall over after their roots are killed and generally fail to recover.

Sclerotinia diseases

These are among the most common decay fungi in the garden. Cottony soft rot, white mould or watery soft rot caused by *Sclerotinia sclerotiorum* is particularly common on camellia, columbine, chrysanthemum, dahlia, hollyhock, narcissus, nicotiniana, paeony, snapdragon, stock, sunflower and many other herbaceous plants, as well as numerous vegetables. Symptoms vary with the part of the host that is affected and with the prevailing environmental conditions. Succulent herbaceous plants initially develop pale brown or dark brown patches at their bases. At the early stages of lesion development, only the stem is affected and the leaves remain virtually symptomless. Occasionally infection starts from a diseased leaf and passes into the stem. In both cases this means that infected plants are easily missed during inspection. In later stages the plants with infected stems die, the foliage becomes wilted, collapses and also appears obviously dead, so infected plants are much more easily spotted. Apart from the unobvious early symptoms on the host, the most visible early sign of infection is usually the white cotton-like ball of mycelium that rapidly develops around the point of infection on the host. Sometimes this profuse mycelium fills the pith cavity of the infected host plant. Often this dense, fluffy mycelium will eventually covers both the surface of the plant and may contain several sclerotia. Sclerotia are hard, black structures that vary in size from two to 10mm in diameter that allow some fungi such as *Sclerotinia* as well as those that cause grey moulds and white rots to survive after the host plant has stopped growing. Together with the mycelium that survives in infected tissues, the sclerotia on or within infected tissues as well as those that fall on the ground are the major source of overwintered inoculum. Sclerotia can survive for at least three years or more in the field, but will not all germinate at the same time.

They are frequently found on the plant surface or deep inside the host tissues that are full of mycelium. Although these are white at first, the hard black sclerotia formed in old infections usually germinate after dormancy to form mycelium, or sometimes disc or cup-shaped apothecia develop and release ascospores into air currents over a period of two to three weeks, particularly during damp weather. Although the sclerotia germinate and produce mycelial strands that infect plants directly through moist soil, airborne ascospores are generally the most important means of spread. Most plants are attacked above and at soil level at any stage of development. Stems and stalks develop sunken, elongated lesions along their entire length in wet weather, these often bear sclerotia. flower infection is common in several hosts, particularly camellia and narcissus. Variable numbers of light tan, small, watery spots appear on the petals, which enlarge by coalescing to cover the whole petal or flower. After the flower is infected it drops and is then eventually covered in white mycelium and decays often after forming sclerotia.

Sclerotinia can be controlled by several cultural practices, as none of the effective chemical treatments are available to gardeners. Professional growers also use steam to sterilize the soil of glasshouses and sometimes outside beds. Gardeners can reduce the effects of disease by growing susceptible plants in well-drained soils, thus avoiding the high air and soil moisture that favour the disease. For the same reason, avoid planting herbaceous plants too close together or letting them get crowded out by weeds. Infected plants should be rogued out and burnt to destroy sclerotia.

Stool miners

The adult chrysanthemum stool miner (*Psila nigracornis*) is a shining-black fly related to the carrot root fly which it quite closely resembles. Two generations are produced each year. Outdoors the first brood emerges between early May/early June. The eggs laid around chrysanthemum plants hatch within a fortnight into the larvae: thin cream-coloured maggots which

Left**: Fungus fly larva in narcissus.**
Above: **Stem rot sclerotia formed at plant stem base.**

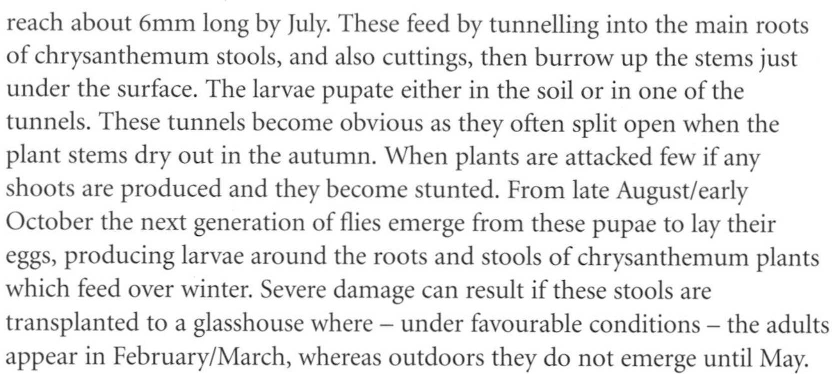

reach about 6mm long by July. These feed by tunnelling into the main roots of chrysanthemum stools, and also cuttings, then burrow up the stems just under the surface. The larvae pupate either in the soil or in one of the tunnels. These tunnels become obvious as they often split open when the plant stems dry out in the autumn. When plants are attacked few if any shoots are produced and they become stunted. From late August/early October the next generation of flies emerge from these pupae to lay their eggs, producing larvae around the roots and stools of chrysanthemum plants which feed over winter. Severe damage can result if these stools are transplanted to a glasshouse where – under favourable conditions – the adults appear in February/March, whereas outdoors they do not emerge until May.

There are a number of insecticides that can be applied to the soil around the base of chrysanthemum plants in early May and early September within a week of planting to protect them from the larvae hatching from eggs and some others that will penetrate the stools and exterminate the larvae already inside.

Springtails

Springtails are minute wingless insects of various colours ranging from white, grey, brown or green, usually just over one to 2mm long which can occur in enormous numbers in and on the soil. Many species can jump by means of a forked structure that is normally held under the abdomen but applies pressure when forced onto the ground. They generally feed on moist, rotting plant debris but they can also be found on the decaying tissue of plants that have suffered some other damage and a few species can injure healthy plants. Some parthenogenetic species do not need to mate before producing eggs, but other species breed in a normal fashion. The eggs that are laid in the soil in batches or singly hatch into minute adult forms without metamorphosis, but they continue to moult even after they have become mature and sexually reproductive. The garden springtail (*Bourlietiella hortensis*) and species of *Onychiurus* nibble into the stems and roots of healthy seedlings. This injury stunts the subsequent growth of the plants which will sometimes wilt if root damage is especially heavy. The aerial roots of orchids can be damaged if springtails infest the damp moss on which they grow. The holes that result from springtails gnawing through the stem or leaves that touch the soil are often the sites for soil-borne pathogens to enter and the rot the plant, often causing it to collapse. Springtails are most abundant where the soil is acid and damp, so liming and proper drainage can help to reduce the severity of infestation. Their numbers also increase in glasshouses due to the milder conditions. Although sterile seed composts should not be infested, the most usual method of control is to incorporate a suitable insecticide into compost in the seed tray or pot, seed bed soil, or rake this into the soil around affected plants.

Swift moth caterpillars

The caterpillars of swift moths also feed voraciously on a wide variety of garden plants. There are two species which are most commonly found destroying plants in gardens. The garden swift moth (*Hepialus lupinula*) is more common on herbaceous, bulbous, corm and bedding plants, whereas the ghost swift moth (*H. humuli*) generally attacks grasses in lawns. Both caterpillars have white bodies and a chestnut-coloured head and are about 6cm long. The garden swift moths have yellowish brown wings but the ghost swift moth is larger and paler. The garden swift moths lay their eggs in flight in May/June near appropriate host plants. The caterpillars that result burrow through the soil until they reach the fine roots of a suitable host plant when they will start to feed, ultimately reaching the thicker roots and tunnelling up the lower stems or into bulbs and corms. The caterpillars pupate in the spring under the soil surface and the life cycle of the garden swift moth is usually completed in a year. The ghost swift moths lay their eggs in the same way in June/July, but the resultant caterpillar takes two years to become a moth.

One of the most effective ways to control swift moth caterpillars is to dig the soil thoroughly before planting susceptible host plants such as aster, campanula, chrysanthemum, coreopsis, delphinium, lily of the valley, lupin, Michaelmas daisy, phlox, dahlia, gladiolus, iris, narcissus and peony. After planting out care should be taken to ensure that the soil between plants is kept cultivated in order to injure the caterpillars and expose them to birds such as starlings that eat them avidly. It also pays to lift herbaceous and bulbous plants occasionally to examine the roots for caterpillars. After these have been removed the plants can be replanted in soil that has been treated with a suitable insecticide. Established plants can be drenched with insecticide solution and bare soil treated during cultivation.

Symphylids

The symphylids are related to the millepedes and centipedes and the only pest of any importance is the glasshouse symphylid, white insect or glasshouse centipede (*Scutigerella immaculata*). These highly active pests are made up of 15 thin white segments about 8mm long with a pair of long antennae and 12 pairs of legs. They might be confused with the wingless insect *Campodea* but this has only three pairs of legs and two tail filaments.

***Top left*: Root fly larva.**

***Left*: Springtail in soil.**

***Above*: Small garden swift moth caterpillar.**

Symphylids lay their eggs in the surface layers of the soil in batches of two to 20 which hatch after a few weeks into forms with six pairs of legs and short antennae, but at each moult extra legs and longer antennae result, until the sixth moult when the adult form is reached. Under favourable conditions the life cycle can take three months but is generally longer. However, breeding continues unceasingly unless temperatures are excessively high or low when the symphylids burrow deep into the soil, often for several metres. Even so, symphylids are most common from late spring to autumn outdoors, but in glasshouses they peak from late winter to early spring.

Large populations of symphylids can occur in glasshouse soils, but they can also be found damaging plants outdoors such as anemone, calendula, chrysanthemum, freesia, primula, smilax, sweet pea and violet. The main damage is caused by the symphylids gnawing away the root system of the plant. When the root hairs and fine roots are lost the plants do not thrive and can become stunted, wilt in hot weather or even die especially if root rot pathogens are able to enter wounds, particularly those on the larger roots.Frequently those leaves that touch the soil are also devoured. Prevention is easier than cure so it is wise to avoid bringing in plants or soil from areas known to be already infested, as once symphylids have been introduced they are impossible to exterminate because they migrate so deeply into the soil that insecticides cannot reach them.

Vine weevils

The initial symptoms of vine weevils are yellowing and wilting of foliage despite plenty of watering. The adult vine weevil, *Otiorhynchus sulcatus*, is a blackish wingless weevil about 8mm long with tiny yellow bristles. It is seldom seen during the day except when its hiding places are disturbed, but can be found at night when plants are illuminated with a torch. The adults are vigorous climbers and crawl over plants. Both the adults and grubs can be devastating to many cultivated plants particularly cyclamen, fuchsia, bedding plants, especially primula, many herbaceous perennials, ferns, alpines, especially saxifrage and most hardy nursery stock such as rhododendron and camellia that produce new roots slowly. There are four stages in the life cycle: adult (beetle), egg, larva (grub) and pupa. There are no males – all the eggs are laid by unfertilised females. Although there is only one generation each year, there is often sufficient overlap so that all stages can be found at any time of the year. The adult weevil feeds at night on the foliage causing characteristic notch-like edges to the leaves from late autumn in glasshouses but later outdoors, usually from late May onwards. Eggs laid in crevices in the soil during the summer from July to September, hatch into creamy-white, brown-headed, C-shaped grubs within weeks. By the following March the larvae, now 10mm long, mature into pupae. The smaller larvae feed on the finer roots, whereas the larger larvae destroy the main roots and corms, causing plants to wilt and die from late summer to spring. Among the plants that are most often attacked by grubs are azalea, begonia, cineraria, coleus, ferns, hydrangea, lewisia, primula, saxifrage and sedum.

Many chemicals have been tried for the control of vine weevils, including many organochlorine insecticides, but without major success as the adults are very resistant. Some success has been reported on woody plants using tree-banding grease which prevents the wingless adults from climbing up to the foliage. Another method for controlling adults is to lay loosely rolled corrugated paper or sacking at the base of damaged plants in order to trap the weevils during the daytime when they hide. The traps should be examined daily and the weevils destroyed. Treating the larval stage is also not easy, as it is necessary to lift and repot plants infested with larvae into fresh treated soil or compost after picking off the grubs and removing the original soil. Alternatively insecticide has to be drenched through the soil, if necessary through holes that have been made in the soil. None of these techniques has been as successful as biological control using an indigenous species of nematode, *Heterorhabditis megidis*, which is associated with a bacterium, to attack the larval stage.

Several products based on this or another nematode, *Steinernema carpocapse*, are available in Britain by mail order. These are formulated in a mud-like carrier which can be stored in the refrigerator but should be applied as soon as possible once the larvae are present. The nematodes need warm soil so that after dilution they can be applied with a watering can with fine rose at any time under glass as long as the temperature does not fall below 12°C, but outdoors they work best when applied from mid-August to mid-October and April to the end of May. One company that markets *Heterorhabditis megidis* say it is also worth applying an insecticide such as pirimiphos-methyl as a pre-treatment from May to June. The nematodes are attracted to the larvae which they penetrate through their body openings and the bacteria that they are associated with them kill the grubs within 48 hours. Once the grubs have been invaded the bacteria grow, helping the nematodes to grow and reproduce. After the grubs die the nematode eggs within them hatch releasing nematodes into the soil to spread the infection to other survivors or to new generations of vine weevils.

Above: **Vine weevil larvae on plant roots.**

Wirestem

Rhizoctonia solani is a very common fungus in most soils because it can survive both by living on plant debris and also as a dormant sclerotium.

It can be seed borne but more usually spreads through the soil as a weft of coarse mycelium without spores that can sometimes be seen with the naked eye. The sporing perfect stage of the fungus is rarely seen on white mycelium in the collar region of the host.

Often the coarse mycelium will eventually congregate into a rough ball on infected plants to form the sclerotia, 2mm diameter, solid blackish-brown structures, which commonly form on the surface of potato tubers, where they look like pieces of hard coal. These sclerotia can remain dormant in soil for long periods and will germinate to produce fresh mycelium when a host seed germinates or a host is planted nearby. This fungus is responsible for attacking the roots and stem bases of a very wide range of plants. Although many different hosts are attacked, several strains of *R. solani* are involved. Often each of these is only associated with quite distinct host species. It will attack both seedlings as well as more mature plants. Among the diverse plants attacked are most annual plants including vegetables and weeds, many perennial ornamentals, turf grasses, shrubs and trees. The symptoms vary depending on the age of the plant affected. In the earliest stages of seedling emergence, the root tip is attacked and destroyed. If the interior of the root or stem base of an older seedling is harder than the surface layers it often remains long after the soft outer tissues have disappeared as a thin flexible filament connecting the root tip and the shoot in a symptom called 'wire-stem'. In more mature plants *Rhizoctonia* produces stem cankers which are brownish, rough, cracked lesions, usually in the crown region, as well as root and stem rots. The lesions may increase in length and may penetrate the woody tissues to extend up the pith into the stem and even the lower leaves, but many attacks are not very serious. In these outbreaks, trivial decay may be seen in addition to the few symptoms that are visible, apart from some obscure injuries on the roots and stem bases. In more severe cases the plant may be completely stunted, the leaves may curl up and discolour yellow or purple. On tubers, bulbs, corms and other fleshy, succulent stems or roots *Rhizoctonia* causes superficial or deeper rotten areas that appear brownish, often extending inwards to the centre of the root or stem. As the rotten areas decompose and dry out they leave behind a sunken area often encrusted with sclerotia. Sometimes these lesions cause a crown rot that stunts and yellows the foliage. In wet weather webs of tan to cream mycelium cover these lesions. Remove, destroy and burn all dead or dying plants as well as any susceptible weeds. Commercial growers treat soil with an appropriate fungicide, generally toxic fumigants that have to be applied by a specialist contractor, but gardeners are not allowed the most effective products so they will find control is difficult, although a well prepared seedbed and other operations that improve drainage and/or nutrient supply often ensure reasonably successful germination in garden soil. Three-year rotations can be carried out but are only of limited use once the fungus is present at high levels, as in the spring where soils are cold and dry. Some authorities suggest shallow planting since this allows the plants to establish more quickly and they emerge faster, thus reducing the chance of infection in the soil.

Wireworms

Wireworms are the larvae of click beetles, in gardens these are generally *Agriotes lineatus* and *Athous haemorrhoidalis.* Both of these adults are slender brownish beetles that have the ability to click audibly by pressing the edges of their thorax and abdomen together and hence flick themselves into the air if they fall or are placed on their backs. As larvae they are firm glossy golden or reddish yellow wireworms about 25mm long. These elongated hard-skinned larvae have a set of powerful jaws which they use to feed on the roots of many plants and to burrow into tubers, corms and rhizomes. They may sometimes be found inside the stems of fleshy plants well above the soil. Click beetles lay their eggs under the soil below the cover of grass or other plants including weeds from May/July to avoid desiccation. After a few weeks the larvae hatch and take four to five years before they burrow deeper into the soil to pupate in an earthen cell in July/August. While the larvae are still developing they show two peaks of feeding in March/May and late summer/autumn. After pupating for a few weeks, the click beetles emerge from the pupae but often remain inside the earthen cell over winter; a few adults emerge from the soil and hibernate among rank vegetation.

It should be possible to distinguish wireworms easily from some other similar soil inhabiting creatures such as millepedes or centipedes which have more legs, the more active larvae of ground beetles and the legless maggots of stiletto flies. All of these except for the millepedes are beneficial predators that feed on plant pests and so should not be killed.

Wireworms are most common after grassed areas have been cultivated. In this situation, an approved insecticide can be incorporated into the soil a few weeks before seeds are sown or plants are transplanted. It is possible to reduce wireworm damage without resorting to chemicals by destroying all potential weed hosts and repeatedly cultivating the soil thoroughly in order to produce a good tilth structure that is dry and firm enough to prevent the eggs from hatching and hence discourage the female click beetle from laying eggs in the first place.

Above: **Wireworm (click beetle larva) feeding on damaged potato tuber.**

(v) Seeds and fruits

Ants

Ants are only a minor problem on fruit which they are attracted to when it is fully ripe and even then they are usually just a nuisance. The most common species in gardens are *Lasius niger* the common black ant and *Lasius flavus* the mound ant. Among the other damage caused by ants is the hoarding of seeds for food. The ants that build mounds can interfere with grass mowing. It is essential to apply insecticide to the nests if ants are to be eradicated. If the nests cannot be discovered it is necessary to apply a bait. In those that are available, the insecticide is formulated with sugar making it attractive to foraging worker ants which carry portions back the nest to feed other ants which are then exterminated. If the queen ant is killed then in a short time the nest will be destroyed.

Beetles

Few species of beetles are common pests of fruit even though there are many beetles in natural environments. Generally beetles are easily recognised and familiar, despite this the larvae of beetles tend to be remarkably dissimilar (see p. 54). One beetle where the adults and larvae eat the seed pods of a range of liliaceous plants is the lily beetle (*Lilioceris lilii*) which also devours the leaves and stems. It is particularly destructive on lily, *Fritillaria* spp., *Nomocharis* spp. and *Polygonum spp.*. Although some insecticides are effective, it is often just as easy to pick off and squash the conspicuous bright scarlet beetles and reddish faeces-covered grubs.

Birds

A few birds are unappreciated because of the damage they cause to displays of autumn berries on shrubs and trees, such as cotoneaster, holly, pyracantha, rowan and stranvaesia. In autumn, these are often promptly stripped bare by throngs of blackbirds, thrushes, jays, starlings and wood pigeons. However sometimes varieties of shrubs with berries of a somewhat unusual colour are occasionally avoided by the winged marauders. Most birds are protected by law but shooting or trapping can be carried out under some circumstances but these should be corroborated by the appropriate authority such as the Ministry of Agriculture.

Otherwise the only way to avoid bird damage is to scare them away by offending their senses of sight, sound or touch. Traditional scarecrows are only effective so long as they are regularly moved about or are capable of rocking in the wind. Other devices depend on strips of aluminium foil that glitter and twist, sometimes showing the colour on the other side as they are blown by the wind. These can be bent to make a noise as well but noisy birdscarers are usually discouraged by neighbours. Black thread strung out between the branches of shrubs will discourage birds from sweeping in to feed but can be rather unsightly. Repellents that are sprayed onto plants have generally met with little real success and it is illegal to use substances that glue the legs of small birds to twigs such as the traditional bird lime used by the birdcatchers of old, but some sticky substances can be used to hinder perching on branches.

Brown rot

Brown rot causes a blossom blight and a fruit rot, it can also cause infections on twigs. Infection often spreads to other pears stored nearby after harvest. Hard black bodies, the sclerotia that form in old fruit and twig infections usually germinate to release spores in the spring. Remove infected fruit and provide suitable ventilation for harvested fruit. Many preharvest fungicide products are used commercially, a few are available to amateur gardeners, but chemical sprays are only partially successful under heavy disease pressure.

Caterpillars

Most caterpillars, the larvae of moths and butterflies, feed voraciously on the foliage of their hosts plants with their strong toothed mandibles, others are stem borers or live on roots in the soil and a few attack fruits. None of the adults are at all injurious as they only feed on nectar and some other liquids through a tubular proboscis. The caterpillars of the delphinium moth (*Polychrysia moneta*), which is a tortrix moth, feed on the seed capsules as well as the leaves, flowers and buds of Delphinium, larkspur and Aconitum which they first tie with silk. Delphinium moth, like most tortrix, caterpillars are hard to treat with insecticide sprays because they stay concealed within their bundles of leaves, so they most easily picked off and burnt. However, if an attack is anticipated many caterpillars can be controlled by a prophylactic spray of insecticide or biological control agent based on the spores of the bacterium, *Bacillus thuringiensis.*

Left: **Pecked and damaged rose hips.**
Top: **Brown rot infected fruit.**
Above: **Large white butterfly caterpillar.**

Mammals

Mammals of various sorts occasionally eat the fruits off plants in the garden but most debark trees, damage plant stems, eat foliage, flowers or buds and uproot plants. Nonetheless, bank voles can strip the fruits off bushes as they readily climb up them. They can cause damage especially during those years when their populations become numerous but can be discouraged if the grass or other plants that grow around their nests is removed exposing the bank voles and their young to attack by predators such as domestic cats and dogs, foxes and weasels.

(vi) Shoots and stems

Aphids

Although most aphids, the familiar greenflies and blackflies, are most commonly found on the foliage or roots (see p.51 and p.113), colonies of several common aphids also cluster on the stems of herbaceous plants in summer or on the branches of the woody plants where many overwinter. During the summer swarms of winged migrants use calm air currents to drift long distances to new herbaceous hosts but in the autumn they migrate to a winter host, usually a woody tree or shrub in the same way. Some of the eggs that they lay there survive until spring in a crack or bud before hatching into winged females. By parthenogenesis, these produce several generations of female aphids, some of which have wings and migrate to the host plant preferred during the summer. Normally this is a cultivated or wild annual or herbaceous plant. More winged and wingless generations of female aphids are produced there during the rest of the summer until winged female aphids migrate back to the winter host to produce wingless female aphids. These wingless female aphids stay within their colonies on their winter host plants where they lay the overwintering eggs after mating with winged males. Nonetheless some species of aphids, often those with wingless males, remain on the herbaceous host plants all year and do not have a winter host. Some species overwinter as nymphs and adults may resume parthenogenic reproduction if conditions are opportune. During each week of part of the summer, a female aphid can produce over 50 young, and by the beginning of the following week these young will also be ready to reproduce. Hence a negligible attack can quickly develop into a severe infestation.

Aphids that feed on stems and shoots of their summer hosts result in typical distortions caused by both feeding by sucking up sap and also because of the toxicity of aphid saliva. Other symptoms result from the virus and other diseases that are also spread when the aphids feed or enter through the feeding scars. Aphids also excrete a tacky, sugary honeydew which spoils the appearance of the plants, and is particularly filthy after sooty moulds have grown on the tiny droplets. Among the most common and damage species that are found on the stems of their herbaceous hosts or the woody overwintering host are the black aphid (*Aphis fabae*) on a many herbaceous plants in summer and on viburnum and the spindle bush in winter, there are others such as the cherry aphid (*Myzus cerasi*) on cherry and various weeds, the peach-potato aphid (*Myzus persicae*) on a wide range of glasshouse and outdoor plants to which it transmits several viruses,

Left: **Harvest mouse with rowan berries.**
Above: **Rose aphid colony on rose leaf.**

the potato aphid (*Macrosiphum euphorbiae*) on various summer hosts and on rose in winter, the rose aphid (*Macrosiphum rosae*) on rose buds, holly, scabious and teasels during the summer and the water-lily aphid (*Rhopalosiphum nymphaeae*) which overwinters on species of *Prunus*. Similar forms of aphid species tend to look alike, so it is often difficult for someone who is not an expert to differentiate easily between them. In any case, aphids are normally controlled in similar ways by the same chemicals. Examine plants thoroughly, especially those about to be purchased so that you can prevent heavy infestations building up by timely treatment. Aphids attract a wide range of natural predators and parasites, these need to be protected against insecticide sprays. Often systemic aphicides have been developed for this purpose, but these should not be applied on the summer hosts during drought conditions unless the plants have been well watered first. One biological control agent, the gall midge (*Aphidioletes aphidomyza*) which lives entirely on an aphid diet, is available commercially. Gall midges are often applied seven days after a systemic aphicide that does not kill predators. Ants should be controlled as they remove and kill the gall midge larvae. In another method of biological control, fish and other pond animals can be encouraged to catch and eat water-lily aphids by submerging the infested plants under the pond water. Aphids, nymphs and eggs on the woody hosts are often controlled by winter wash applications but have less effect against woolly aphids due to their protective covering.

Woolly aphids

Although it is a distinctive aphid, the woolly aphid (*Eriosoma lanigerum*) found on apples and some other rosaceous shrubs, often on pruning wounds and grafts which swell and split as a result, behaves in similar ways to other aphids (see above and p. 51). However, woolly aphids are not easy to wet without additional wetting agent and so are often not well controlled by the winter washes applied to their woody hosts.

Beetles

Several beetles damage the branches and trunks of woody plants. There are several different species of ambrosia beetles, also called shot-hole borers which attack deciduous and coniferous trees. These beetles are often found on trees that are already infected by diseases, particularly root infecting fungi, but healthy trees can also be colonised. One of the most common shot-hole borers found in the garden is which bores into the heart-wood of several fruit and ornamental trees. This tiny beetle (3mm long) is often found on chestnut, oak, ornamental cherry and sycamore, where it can be detected by finding the tunnels which are evident because of the wood-dust that is released from the holes that pepper the branches and trunk of infested trees. The foliage on heavily infested branches eventually often shows signs of wilting. If an infested branch is broken open, the curved legless white grubs with brown heads can be seen in the tunnels. These larvae do not tunnel through the wood but feed on the fungal spores that are produced on their walls. These fungi originate from spores carried by the adults which create the tunnels before they lay their eggs in them. These become the larvae which hatch, live and pupate in them. In spring the male beetles emerge before the females, which search for new sites to excavate tunnels either on a fresh branch or new host tree. The best way to control ambrosia beetles is to saw off and burn all the branches that are infested, making sure that no tunnels run into the surface of the cut on the tree. Several other treatments have been tried including painting the shot-holes with a tar oil wash in the spring as this forces the beetles out of their tunnels to their death. Other insecticides have also been used in a similar way. A number of bark beetles also cause problems in gardens. Although they are insignificant in size (1.5-4.5mm long), in many areas the presence of populations of elm bark beetles (*Scolytus scolytus* and *S. multistriatus*) is evident as they act as vectors for Dutch elm disease, some aggressive forms or species of which incite a highly conspicuous vascular wilt but other non-aggressive forms are less obvious. The other bark beetles such as the fruit bark beetles (*Scolytus rugulosus* and *S. mali*) which attack hawthorn and ornamental cherry cause only die-back of the branches, but ultimately this can weaken the tree to the point of death. Many of the trees that are infested are already diseased and have a damaged root system, but others seem quite sound before they are attacked. The damage is caused by the adults and larvae producing intricate tunnels beneath the bark and above the wood. The female beetle bores under the bark to lay eggs in recesses in the mother tunnel that she creates. The legless larvae that hatch have curved white bodies with dark brown heads. Until they pupate, the larvae burrow away from this central tunnel in a radial configuration of tunnels that can be clearly seen if the bark is removed. The complex tunnels produced in this way have such distinctive patterns that they are characteristic of each species of bark beetle, and have earned them the name engraver beetles. The position of the tunnels is revealed as the adult beetles leave exit holes behind when they emerge from the bark. In the absence of effective chemical control, it is very difficult to eradicate bark beetles other than by digging out heavily infested trees or destroying branches by burning them. If large trees are infested the bark colonised by the beetle larvae can be stripped off,

Left: **Woolly aphid colony on twig.**

but this is not very wise as it allows fungi to decay. One method of avoiding the problem in the first place is to make sure that trees are properly planted so that they can continue to grow vigorously, particularly if fertilizer is applied. In order to provide the best conditions for establishment, it is essential to avoid planting trees in soil that is either too wet or too dry, nor too close to the stumps of trees that may have died as the result of a root disease, such as honey fungus root rot. Other pests should be controlled as they will also weaken the trees and hence may make any infestation by bark beetles more serious.

Butt or foot rot of pine

Heterobasidion annosum causes the wood of its hosts to decay without killing the bark. It mainly attacks pine and other coniferous trees but some broad-leaved trees are also susceptible including species of *Sorbus* such as whitebeam. Diagnosis depends on finding the characteristic bracket fungus fruiting body which forms at the soil level. Butt rot should be suspected if there is root contact with a tree that has already died. The first sign of butt rot is usually when the foliage of the tree is less dense than expected with a number of dead twigs and branches, or if the tree produces or loses its leaves later than usual. In later stages the tree dies or falls over and produces the fruiting body. Dead trees and their should be removed as they can spread the disease by root contact.

Canker

Nectria galligena forms cankers on twigs, branches, large limbs and trunks of apple, pear and a large number of other hosts. Dark water-soaked patches bordered by healthy bark. The rim around the infected area cracks, the wound tissue beneath becomes infected and dies leaving a canker. As the canker is perennial, each year it enlarges by another concentric layer of dead wound callus. *Nectria* readily girdles smaller twigs so these are more readily killed than branches. Subsequent snapping off of weakened twigs and branches results in crown dieback or even the premature death of the tree, where a canker has damaged the trunk. Enlargement of a canker depends on the strain of *Nectria*, the variety of the tree and its host reaction, local environmental conditions, the time of year when wounded, the size of branches and date they were infected. Trees wounded in autumn and winter are more often infected than those injured in late spring or early summer when infection is restricted by more active cork production. *Nectria* can cause eye-rot even on fruit still on the tree *Nectria* continues to form

fruiting bodies and sporulates for several years on dead wood so prune out and burn young twig and branch cankers and, if very severe, infected trees. Fertilize, water and prune healthy trees early 'to stimulate vigorous growth but regulate the nitrogen supply. A few fungicides are available for garden use as preventive sprays and paints for sealing wounds.

Capsids

Three main species of capsids, the common green capsid (*Lygocoris pabulinus*), the potato capsid (*Calocoris norvegicus*) and the tarnished plant bug or bishop bug (*Lygnus rugulipennis*) can injure plant stems as well as leaves and flower buds. Capsids often drop to the ground and run off fast to avoid capture. Tiny brown spots are left on the shoots which have a puncture in the centre marking where they fed. The common green capsid is found on many flowering trees and shrubs, as well as many herbaceous plants and an assortment of weeds. The somewhat similar potato capsid is also found on p.57. The tarnished plant bug is found on herbaceous and annual plants including weeds where it overwinters inside hollow plant stems and decaying rank vegetation.

It is essential to inspect plant stems closely for the first signs of puncturing by capsids as some insecticides can limit serious injury providing they are applied sufficiently early to the plants and the ground beneath them. Any weeds and decaying rank vegetation that could feed and hide capsids should also be gathered up and destroyed.

Caterpillars

A number of caterpillars are stem borers in either herbaceous or woody hosts. The former 'include the pinkish-brown caterpillar of the frosted orange moth (*Gortyna flavag*) which tunnel into the stems of chrysanthemum, dahlia, foxglove, hollyhock, lupin and marigold. Outbreaks of the caterpillar are rarely serious as only a few plants are usually attacked, but these often wilt and die. The brownish yellow moth is active laying eggs at the base of its food plants from late summer to autumn. The eggs overwinter in the plant and the caterpillars that hatch in spring pupate inside their tunnels near the base of the plant in July. The most effective method of control is to carefully cut off affected stems and to crush the caterpillars and pupae inside. The rosy rustic moth (*Hydraecia micacea*) is similar in size and related to the frosted orange moth, has a similar life cycle and is found at a similar time. Its caterpillar causes similar damage by tunnelling down to the roots of herbaceous hosts, particularly antirrhinum,

Top left: **Shot hole borer beetle in wood.**
Left: **Butt rot fruiting stage on pine tree.**
Above: **Canker on pear tree & leopard moth**

chrysanthemum, dahlia, iris and sunflower, but pupates in the soil. The rosy rustic moth can be distinguished as it is rosy pink with brown stripes and the caterpillar is a dull pink with brown spots. Again the best way to control the spread of the infestation in subsequent years is to search for all the infested stems and either burn them or squash the caterpillars inside. The goat moth (*Cossus cossus*) has dark red caterpillars with pinkish-yellow sides which bore into the wood of the trunk and larger branches of several ornamental and forest trees including ash, birch and willow. The goat moth is rather a large dull brown moth which lays its eggs in midsummer in wounds or cracks in the bark. The caterpillars spend the next three or four years feeding on wood inside the tree, usually their presence is revealed by the sawdust-like debris and faeces that fall from the holes that they make through the bark. The caterpillars pupate in a cocoon in the tunnel near the bark. If the entrance holes are discovered early enough, the caterpillars can be removed by impaling them on wires that are poked into the tunnels or by applying a suitable insecticide within them. The yellowish-white, black spotted caterpillar of the leopard moth (*Zeuzera pyrina*) is also found in tunnels inside the wood of the branches of a variety of trees, shrubs and climbers including hawthorn, honeysuckle, laurel, lilac, ornamental cherry, apple and rhododendron. The moth is so-called because it is white with dark spots. The damage is less than that caused by the goat moth, even though the leopard moth is more common, because infestations rarely affect the trunks. Again the presence of infestations can be detected by looking for the tunnel entrances with their tell-tale accumulations of sawdust-like debris and excreta. Another early symptom is the onset of withering of the leaves on affected branches. The main difference in behaviour between the caterpillars of the leopard moth and those of the goat moth is that the former overwinter under the bark in their first year before tunnelling into the bark and that they rarely colonise branches which are more than 10cm in diameter. Any branches that show signs of extensive infestation should be cut off and burnt but early stages in colonisation can be controlled in the same way as for goat moths. The pine shoot moth (*Rhyacionia buoliana*) is a small reddish-orange moth with silver bars that is found in young pine plantations in July/August where the reddish-brown caterpillar tunnels into young shoots after it hatches from eggs laid near the terminal whorl of needles. The exudation of resin from the side buds surrounding the terminal whorl which are colonised is an early sign of infection. The caterpillar overwinters inside a hollowed out side bud before tunnelling on to infest the shoot which wilts, collapses and often dies after being penetrated. This is damaging to ornamental pines, but more serious injury results whenever the leading shoot recovers as the subsequent growth of the tree is often either distorted or forked if a side shoot becomes dominant. The caterpillars pupate in the tunnels. One of the best ways to control pine shoot moths is to detect the early stages and cut off any shoots that are infested, but on large plantations insecticides are often applied.

Crown galls

Crown gall is frequently found on many herbaceous and woody plants where it is the commonest cause of galls at soil level. The galls that are produced are simply swollen masses of cells which the bacterium, *Agrobacterium tumefaciens*, has caused to both enlarge and proliferate sometimes to a fairly spherical mass over one metre in diameter. On trees the crown galls are hard, but on herbaceous plants they are much less firm. Although most of the galls are found at soil level they can be found up the stems of infected plants and occasionally galls can be seen on the leaves. Some other bacteria and can cause similar galls, for example, maize smut but in this case crown gall cannot be implicated as it does not infect cereals and other grasses. Some other plants have galls and burrs that are not caused by a pathogen, like the burr knots on apples. In most cases there is no reason to control crown call as the host generally does not become weakened sufficiently to warrant treatment. However, if necessary the galls can be cut out and burned. It is also wise not to plant other susceptible trees, shrubs or herbaceous plants in the area known to be infected which should be grassed over or used for other non-susceptible monocotyledonous plants.

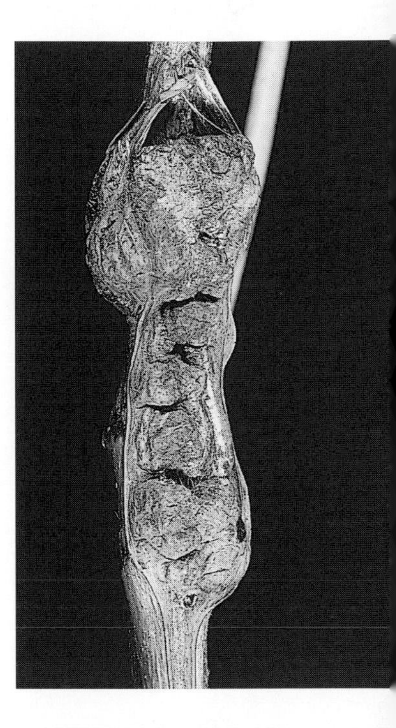

Above: Crown galls on stem.

Dutch elm disease

The epidemics of outbreaks of Dutch elm disease which have killed millions of elm trees in Britain are due to a new highly pathogenic species *Ophiostoma novo-ulmi* which is more virulent than the species that was present earlier *O. ulmi.* This species was introduced into England via North America in the 1960s and has replaced *O. ulmi.* It attacks all European species of elm, as well as *Zelkova*, but this tree is rarely killed. Some resistant species from eastern Asia have been used in breeding programmes but it is not yet known if any of these resistant hybrids will resemble the taller European species. The symptoms of this disease are the appearance of patches of leaves from individual branches becoming wilted or yellowed then turning brown and falling off. The tips of affected twigs become hooked over. In later stages of disease trees may not produce any leaves, or if they are formed they quickly die. The symptoms on the infected trees spread until the tree is dead, this may take a few weeks or several years.

The fungus grows under the bark of dying trees in the breeding galleries of *Scolytus* elm bark beetles. In spring and summer the newly emerged beetles fly off to feed on the sap in the twigs of the crown of neighbouring trees. As they are covered in spores from the breeding chamber, they act as a vector when they feed on the twig crutches, the fungus is then able to spread down the tree as a yeast-like form in the xylem which contains water. As this form passes through the tree it produces toxic substances which provoke a response in the wood which becomes stained by longitudinal streaks. Some caution is required as similar symptoms are caused by *Verticillium* wilt. Once a tree is infected it can spread the disease by root to root spread. Also many hedgerow trees grow as suckers from an original tree. In both cases disease spread is rapid. At present there is no really effective control, although in the initial epidemics contractors applied fungicide into trees under pressure. To be effective the affected branches should be removed or the tree should be pollarded. Also the trees must be at the earliest stages of disease and isolated from other trees. In practice few elms were considered worth the expense involved or did not meet the criteria and there are few elms that were saved. Sometimes the isolation from other elm populations was fortuitous, for example those along the south coast of England, particularly those mature elms in Brighton.

Mammals

Several mammals can cause considerable damage in the garden by eating plant stems, leaves flowers and buds, as well as uprooting plants in the process. Grey squirrels, for example often debark trees such as sycamores to lick up their sap. Unless they are shot or trapped there are few effective methods of control especially if neighbours are inadvertently feeding them with food put out for birds.

Rabbits and hares graze on grass and other foliage but gnaw the bark off trees and shrubs in the winter, often killing them in the process. The most usual way of controlling rabbits and hares is to exclude them by making the garden secure inside a 1.5m high wire fence with the bottom 15cm splayed outwards and buried and wire covered gates with a strong sill, that close over buried wire netting. The mesh should be no bigger than 25–30mm, and it is wise to concrete the buried netting wherever possible. All fences can be breached so the trunks of susceptible trees and shrubs can also be protected by wire netting or other tree guards. These can also protect trees against cats which can damage them while sharpening their claws. If you live in farmland or any other area infested by large numbers of rabbits it is worth reminding the landowners concerned that they have an obligation

Left: **Rabbit damage to hawthorn trunk.**

under law to control the rabbits on their land. Rats and mice often gnaw the stems of plants as well as causing considerable damage in other ways. There are a number of baits which are usually applied inside drain pipes to avoid the pellets being eaten by pets. These baits are frequently applied by local council employees who will also destroy wasps nests.

Voles are inconspicuous most years, but occasionally their populations explode in numbers and they become a serious pest. As well as gnawing the bark off the lower parts of trees and shrubs they also bite through the lower stems of a number of herbaceous plants including chrysanthemums, often shearing off the roots. This damage is especially severe if grass or other herbage is allowed to grow around the plants as field voles are usually afraid to show themselves in the open and make runs through rank vegetation from their nests. The bank vole is less shy and will even climb a short distance to collect berries, buds and shoots. Although traps and baits are used, nonetheless the best methods to foil voles are to mow the grass low around trees and shrubs and to make sure that there are no straw or other bulky mulches that can conceal the voles. It is also sensible to guard trees and shrubs with 9.5mm mesh wire netting the base of which is buried in the soil surrounding them.

Sawflies

Most sawfly larvae that attack herbaceous plants feed on leaves (see p. 000) but the larvae of several species are important pests of stems. Sawfly larvae can be recognised if a stem is cut open because although they resemble the caterpillars of butterflies and moths, in addition to the six pair of legs on the thorax, they have six or eight pairs of prolegs instead of the five found in butterfly and moth caterpillars. Although the banded rose sawfly (*Allantus cinctus*) initially feeds on leaves, the larvae burrow into the stems to pupate. The large rose sawfly (*Argeochropus*) is another species that damages the young shoots and flower stalks on roses which become darker and distort where the eggs are laid in rows of slits. Rose bushes can become weakened by heavy infestations. Although it is fairly easy to control most sawfly infestations with insecticides, those in the stems are very effectively protected and so it may be more practical to prune out the affected stems and burn them if there are not too many.

Scale insects

Adult scale insects may not be noticed as the females and young males are often camouflaged by the scale that shields their wingless bodies, and they do not move like other insects, generally remaining firmly attached to their host plants. In addition the adult winged males, if they occur, and the young are also so small that they are often overlooked. As a result a population of scale insects can soon build up on their plant host which feed on its sap causing considerable loss of growth by stunting and yellowing of its foliage. The appearance of the plant is also usually severely blemished by the sooty moulds that grow on the honeydew that the scale insects excrete. In many species reproduction is by parthenogenesis, but where they occur the minute adult winged males do not feed but exist only as long as it takes them to mate with the scaled females. In some species the eggs are retained under the scale of the female and held in place on the plant surface even though she dies before they hatch. In some other species the eggs are protected under a waxy wool-like material instead. The eggs hatch into young nymphs known as crawlers leave the protection of the dead scale or wool and migrate until they find a suitable place to feed as an adult. Although the female retains its scale and stays in the same site until death, the mature males leave their scales to find a mate. There are a number of scale insects that can be found outdoors, these include the beech scale (*Cryptococcus fagi*), the brown scale (*Parthenolecanium corni*), the horse-chestnut scale (*Pulvinaria regalis*), the juniper scales (*Carulaspis juniperi* and *C. minima*), the mussel scale (*Lepidosaphes ulmi*), nut scale (*Eulecanium tiliae*), the scurfy scale (*Aulacaspis rosae*), the willow scale (*Chionaspis salicis*) and the yew scale (*Parthenolecanium pomeranicum*). The female beech scale insects are conspicuous even though they are less than 1mm long, as they have convex yellow scales that are covered in woolly wax which stands out against the grey bark. Copper beech is especially susceptible. The brown scale is commonly found on the bark of a variety of shrubs of different kinds. When it occurs under glass there may be two or three generations each year, but outdoors the crawlers that hatch in mid-summer, overwinter on the younger shoots until establishing themselves in their adult sites the following spring. Brown-scaled females of the horse-chestnut scale produce a woolly wax around their eggs. As well as horse-chestnut, this scale insect can be found on elm, holly, ivy, magnolia and maple. Both juniper scales infest cypress, juniper and thuja in great numbers causing distortion and discoloration. The scales of the female mussel scale resemble tiny grey-brown mussel shells about 3mm long, and protect the eggs overwinter until they hatch early the following summer. As well as crab-apple, *Pyrus* spp. and other fruit trees, dense populations of mussel scales often encrust the bark of box, ceanothus, chaenomeles, cotoneaster, hawthorn, heather and rose. Nut scales can be confused with brown scales but are most frequent only on ceanothus, hawthorn, hornbeam and

pyracantha. The dirty white scurfy scales infest the shoots and stems of roses which not only makes them unattractive but stunts their growth. The whitish willow scale attacks many other shrubs including broom, ceanothus, flowering currant, lilac, privet, spindle and winter jasmine in addition to willows. The yew scale resembles the brown scale but is confined to yew. Infested yew trees can be defoliate and are often covered in sooty moulds which develop on its honeydew. Although some of the scale insects mentioned above are also found in glasshouses and can thrive there, producing more generations a year than outside, a number of scale insects are more commonly found only under glass yet can sometimes be found in sheltered gardens. Among the latter are the cushion scale (*Chloropulvinaria floccifera*), cymbidium scale (*Lepidosaphes machili*), fern scale (Pinnaspis aspidistri), fluted scale (*Icerya purchasi*), hemispherical scale (*Saissetia coffei*), oleander scale (*Aspidiotus hederae*), orchid scale (*Diaspis boisduvalii*) and soft scale (*Coccus hesperidium*). The cushion scale, cushion scalecymbidium scale and the orchid scale are all found on orchids but the oval yellowish cushion scale can be distinguished as it produces a long white egg sac. The cymbidium scale resembles the mussel scale but is only found on cymbidium orchids. The round flat translucent female orchid scales are found on several genera of orchids and palms. The young male scales which are often found in different clusters can easily be distinguished as they are smaller and covered in a waxy felt. The female fern scales found on aspidistra, fern and palm foliage and stems are mussel-shaped whereas the immature males have long whitish fluted scales. The oval orange-brown fluted scale insects are a major pest of *Citrus* and also affect acacia, cytisus and mimosa. The scales of the females are thrust upwards by the globular white wax covered egg-sac. The adult 2.5mm diameter hemispherical scales are more or less circular and convex, varying in from russet to dark blackish-brown in colour but the immature stages are lighter brown. They are found on many hosts including begonia, carnation, clerodendrum, croton, ferns, figs, oleander, ornamental asparagus and stephanotis. The whitish female oleander scales are flat and rounded with a central yellow spot and found on many herbaceous plants, palms and woody shrubs, often including *Aucuba japonica* outdoors. Soft scale is a very flattened yellowish-brown 4mm long scale insect that is common both under glass and outdoors in sheltered gardens as it infests a very wide host range, usually on the undersides of their leaves in layers along the midribs.

A few insecticides are recommended as sprays, but the strong impervious scale that protects the females and their eggs makes scale insects difficult to control, apart from the tar oil sprays to dormant woody plants. It is therefore sensible to dislodge as many of them as possible either by

Left: **Brown scale, one adult upturned to show young living underneath.**

scraping them off or dabbing them off with a cotton-wool bud soaked in an appropriate dilute insecticide. As chemical sprays are only partially effective it is wise to inspect all new plants for scales, and as a precaution quarantine them in an isolated place for a few weeks. If they are found to be infested they should either be treated or destroyed.

Wasps

Although to most people, the common wasp (*Vespula vulgaris*) and the German wasp (*V. germanica*) are more familiarly associated with the buzzing pillagers with painful stings which plague picnickers and bother barbecues, these often cause damage by gnawing through the stems of dahlias and also ripening fruit. Another source of food are the insects and their larvae that they catch and eat. Among these are many pest species. So wasps are not entirely a pest themselves but most gardeners consider they can live without their help. Most of the conspicuous yellow and black wasps that are troublesome in summer, are the sterile female workers but the larger queens are frequent earlier in the year. The latter are the survivors of nests that were active the previous year and are migrating to find a suitable hole either in the ground or in a building, often under the tiles or inside the wall cavity where they will nest and establish their own colony. The other five species in the genus prefer hollow trees or the twigs of bushes. The papery nests are hollow spheres constructed from a form of *papier maché*, pulped wood fragments chewed off convenient wooden objects such as fence-posts or dead trees softened with saliva. Inside this fragile globe is eventually filled with seven or eight tiers of hexagonal cells containing the eggs, larvae and pupae which are attended by several thousand workers.

Although many householders and gardeners trap and drown wasps in glass bottles and jars partly filled with sugary water mixed with detergent, this is not an effective way to reduce wasp populations. Professional pest control workers hunt out the nest and then destroy the whole colony by applying a quick acting insecticide in the evening when most worker wasps have returned for the night. Usually the insecticide is applied as a dust around the outside of the nest hole so it is taken inside on the wings and bodies of the workers as they enter, and kills those inside. Other methods include the use of insecticide paints or smokes, particularly the latter if the nest is in the branches or twigs of bushes or trees. Whenever possible the nest should be removed and burnt to kill the eggs, grubs and pupae.

Witches' brooms

Witches' brooms are the result of abnormal twig proliferation, are often associated with species of *Taphrina*. They are usually found on birch, hornbeam and flowering cherries which are infected by *T. betulina*, *T. carpini* and *T. wiesneri* respectively. Other witches' brooms may have other causes. In the past, witches brooms have been thought to have been caused by the mites that are found living in them, but it is now known that this is not the case. Witches' brooms are found on otherwise normal trees but the dense mass of proliferating twigs tend to come into leaf earlier and lose their leaves earlier than the rest of the tree. The leaves on the broom are stunted, malformed and swollen along their veins. Although the twigs are thin but their bases are thicker than usual and on hornbeam and flowering cherry they usually hang down from the affected branches. The mycelium in the buds survives over winter. During the summer the leaves on the witches' broom that are infected become covered in a white bloom as the ascospores are released. Some of these become blown though the air to new hosts where they germinate to form yeast-like cells which proliferate on the surface of the host. Some penetrate the buds and cause new witches' brooms to become initiated. While witches' brooms may be a curiosity on birch and hornbeam, the lack of flowering on ornamental cherries is regarded as a detrimental side effect, especially as they can become very large. These can be pruned off the tree, this should be done during the summer to avoid infection by bacterial canker and other fungal diseases.

Left: Common wasp - nest exterior wall and inner chamber.

INDEX OF MAJOR PLANT HOSTS OF PESTS AND DISEASES IN THE GARDEN

Left: **Witches broom on silver birch.**